The Architect's Guide to ENERGY CONSERVATION

Realistic Energy Planning for Buildings

Seymour Jarmul, AIA

Partner, Jarmul, Brizee, and Levitt,
Architects and Planners
President, Energy Strategies Inc.

Formerly Adjunct Professor of Environmental
Control Systems at Syracuse University School
of Architecture

McGraw-Hill Book Company

New York, St. Louis, San Francisco, Auckland, Bogotá, Hamburg, Johannesburg, London, Madrid, Mexico, Montreal, New Delhi, Panama, Paris, São Paulo, Singapore, Sydney, Toronto

Library of Congress Cataloging in Publication Data
Jarmul, Seymour.
The architect's guide to energy conservation.

Bibliography: p.
Includes index.
1. Architecture and energy conservation. I. Title.
NA2542.3.J37 729 79-19539
ISBN 0-07-032296-1

1234567890 KPKP 89876543210

The editors for this book were Jeremy Robinson and Christine M. Ulwick, the designer was Mark E. Safran, and the production supervisor was Paul A. Malchow. It was set in Memphis Light by University Graphics, Inc.

Printed and bound by The Kingsport Press.

Contents

Preface

Since the energy crisis became a major factor in the U.S. economy, architects have been flooded with energy conservation advice. It descends from government agencies, from the marketing divisions of our industries, from the literature of professional and trade organizations, and from the research laboratories and seminars of our academic community.

The Architect's Guide to Energy Conservation puts this present outpouring into perspective and provides a basis for evaluating future developments. It brings together information that previously was widely scattered and provides a synthesis that forms a practical working tool for architects, so that they can design buildings within which energy conservation can be effectively and economically achieved.

The book examines ways of achieving energy conservation with the resources currently available to today's practitioner. Procedures are outlined for applying conservation techniques to millions of existing structures as well as for incorporating conservation into new construction.

Energy conservation procedures that increase the cost of buildings are likely to be adopted only if the resultant savings pay off the investment rapidly. The text emphasizes those conservation techniques that fit within realistic construction budgets and are capable of being installed and operated by today's workforce. It provides simplified methods of cost/benefit analyses for evaluating different options and suggests those targets most likely to be profitable when the budget is limited.

In dealing with complex structures, architects have a basic understanding of the mechanical equipment commonly used, but rarely get involved in the convolutions of intricate equipment calculation and design, since these engineering minutiae are handled by appropriate mechanical and other consultants. Accordingly, this book avoids those levels of detailed engineering that will continue to be performed by the appropriate specialists.

Included are a series of case studies of both new and existing buildings. They describe the basis for the design, the cost of the work, and the resultant energy savings.

The text concludes with a consideration of the ways to fund conservation work in both the private and the public sectors and indicates those approaches likely to prove most fruitful.

Although it was written primarily for the architectural practitioner, this text will assist architectural students in understanding the effects of the energy crisis on architecture and increase their professional competence once they leave school.

I am indebted to my architectural partners, Bernard N. Brizee and Abraham D. Levitt, for encouraging my work in the energy field and for graciously providing me the time to pursue these interests.

I appreciate the hospitality and competence of my colleagues at the Syracuse School of Architecture who introduced me to many energy activities of the architectural academic community.

I wish to thank the partners of Goldman, Sokolow, and Copeland, Consulting Engineers, for reviewing the text.

Finally, to my wife Lore and my friend Joe Fischer—my deepest gratitude for their support and wisdom, without which this book would never have happened.

Seymour Jarmul

The Architect's Guide to ENERGY CONSERVATION

Chapter 1

Energy Consumption in Buildings

The energy consumed by a building is the result of the energy needs of the structure and the efficiency with which those needs are satisfied. Energy conservation aims at both reducing basic demand by cutting a building's energy appetite and improving the efficiency of the energy supply system by eliminating waste.

Traditionally the energy demands of a structure have resulted from the actions of those who, be they owners or architects, have controlled the programming, design, and operation of the structure. Similarly, the satisfaction of those demands primarily has been the responsibility of the mechanical engineer.

Under the pressures of the energy crisis the lines separating these participants are beginning to dissolve. No longer can the architect formulate a building's design and then ask the mechanical engineer to "make it work." No longer can the mechanical engineer excuse an inefficient design by rationalizing that the physical space within which one had to fit equipment permitted only a single approach.

The energy demands of a structure are a function of (1) its design, (2) the environment in which it is located, and (3) the way in which it is operated.

A homeowner is primarily concerned with replacing the heat which escapes in winter through the building envelope, or in counteracting the heat which penetrates the envelope in summer. In large part this is an envelope problem. With more complex buildings, however, the effect of the envelope is but a fraction of the total consumption. The lights or computers in an office building may throw off so much heat that the building may have to be cooled even though outside temperatures are below freezing; and in an industrial plant process equipment may completely outweigh the impact of the building skin.

In effect, there are no simplistic solutions which apply to all situations. Before one can determine what steps can be taken to conserve energy, it is necessary to predict how energy will be consumed in a new structure or to document how it is being consumed in an existing one.

PRIORITIES

Nationwide the systems which consume the most energy, in order of magnitude, are heating and ventilating, lighting, air conditioning (cooling) and ventilating, equipment and processes, and domestic hot water. However, the relative order of magnitude of energy use among the first three systems will change, depending on the climate; the building construction; use and mode of operation; and the type, control, and efficiency of the mechanical and electric equipment.

The amount of energy required for domestic hot water is significant in hospitals, housing, and athletic or cooking facilities in schools and colleges. In northern housing the amount of energy needed to heat water is second only to space heating, and to air conditioning in the south. In hospitals the amount of energy required to heat hot water may exceed the amount of energy needed for lighting.

Religious buildings and public halls, which frequently include meeting rooms, offices, and school facilities, are most likely to consume energy in the same pattern as office buildings in the same geographic location—but in smaller quantities per square foot of floor area.

In those retail stores with high levels of general illumination and display lighting, and/or a large number of commercial refrigeration units, electricity consumes the greatest amount of energy.

In climatic zones with mild winters (below 2500 degree-days), the seasonal cooling load may be larger than the seasonal heating load and may even consume more energy, depending on the respective efficiencies of each system. In office buildings, schools, and retail stores in this zone, the electric load for lighting, which is relatively independent of climate, may exceed either heating or cooling. Buildings used for only a few hours per week, however, may consume more energy for heating unless indoor temperatures are set back during unoccupied periods and boiler or furnace efficiencies are high.

In cold climates (6000 degree-days and above) heating usually consumes the greatest amount of energy per year in office buildings and schools, with lighting and cooling next. For retail stores in that zone, the most likely order of energy use is lighting-heating-cooling, or lighting-cooling-heating. Generally, heating consumes the most energy for religious buildings or other buildings used for only a few hours per week in this zone (in fact, in most climatic zones above 3000 degree-days), with lighting and cooling following in that order.

In middle-range climates (2500 to 6000 degree-days), the magnitude of

Table 1-1 Comparative energy use by system

		Heating and Ventilation	Cooling and Ventilation	Lighting	Power and Process	Domestic Hot Water
Schools	A	4	3	1	5	2
	B	1	4	2	5	3
	C	1	4	2	5	3
Colleges	A	5	2	1	4	3
	B	1	3	2	5	4
	C	1	5	2	4	3
Office buildings	A	3	1	2	4	5
	B	1	3	2	4	5
	C	1	3	2	4	5
Commercial stores	A	3	1	2	4	5
	B	2	3	1	4	5
	C	1	3	2	4	5
Religious buildings	A	3	2	1	4	5
	B	1	3	2	4	5
	C	1	3	2	4	5
Hospitals	A	4	1	2	5	3
	B	1	3	4	5	2
	C	1	5	3	4	2

Climatic zone A: Fewer than 2500 degree-days
Climatic zone B: 2500 to 5500 degree-days
Climatic zone C: 5500 to 9500 degree-days

Source: "Guidelines for Saving Energy in Existing Buildings, Federal Energy Administration, 1975.

energy use by systems largely depends on the type of mechanical and electric systems and the characteristics of the building structure in which they are installed. The energy required for industrial buildings exclusive of process loads is generally similar in all zones to that of commercial buildings.

Table 1-1 rates the systems by buildings and climates in the general order of annual energy use, with 1 the greatest and 5 the least. However, each building must be analyzed individually to determine its actual annual consumption by system.

ENERGY CONSUMPTION MEASUREMENT

Energy consumption can be expressed in different ways. Most owners think of it in terms of dollars spent each year. Thus, the yearly cost of energy for various hospital sizes might be expressed as shown in Fig. 1-1.

Others think of energy consumption as a percentage of their total operating budget. Thus the impact of the Organization of Petroleum-Exporting Countries

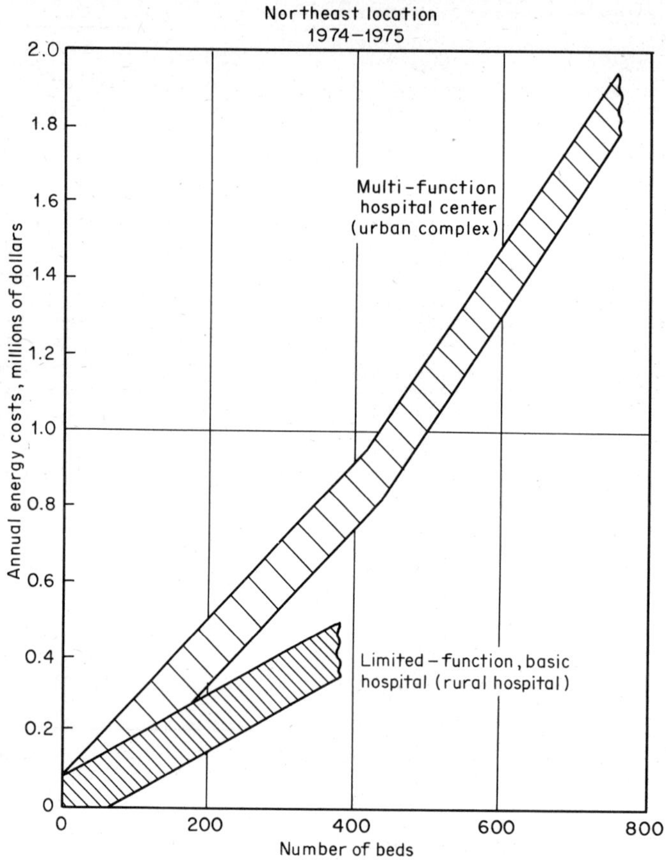

FIG. 1-1 Annual energy cost for hospitals. *("Practical Energy Management in Health Care Institutions." Copyright 1977 by Blue Cross of Greater Philadelphia.)*

(OPEC) crisis of 1973–1974 on New York City's housing is expressed in Table 1-2 in terms of fuel costs as a percentage of total rent.

But budgets and dollars, while significant as the bottom-line objective for most building operators, are not sufficient in themselves to provide the working data needed for effective energy management.

Table 1-2 Impact of 1973–1974 fuel oil cost increases on the economics of New York City's rent-controlled housing

Item		Rent-Controlled Buildings			
Type of fuel oil used		Nos. 2 and 4	Nos. 2 and 4	No. 6	No. 6
Type of building		Elevator	Walk-up	Elevator	Walk-up
Number of buildings		28	135	186	69
Fuel costs as percentage of total rent:	1973	10.0	15.0	8.3	12.5
	1974	19.0	25.9	16.9	23.6

Conversion to British Thermal Units

In actual conservation practice, working data are increasingly expressed in terms of the British thermal unit (Btu), which is the quantity of heat required to raise the temperature of one pound of water one degree Fahrenheit. Thus even though electricity is billed as kilowatthours (kWh) , or oil in gallons, it is helpful to convert all energy consumption into the common language of the British thermal unit.

This text has been written using the U.S. Customary System measurements (e.g., feet, inches, and British thermal units). When the United States finally adopts the metric system, the SI (Systeme Internationale) units, then Tables 1-3 and 1-4 will be helpful in conversions from one to the other.

Accordingly, 1000 gallons (gal) of no. 2 oil = 140 million Btu (1000 gal × 0.14 × 10^6, or 140 × 10^6 Btu).

Table 1-3 Energy conversion factors

Btu Conversion Calculations		
Annual energy consumption	Conversion factor	Total Btu/yr
Electricity ______ kWh	× 0.003412	= ______ 10^6 Btu/yr
Purchased steam ______ 10^3 lb	× 0.97	= ______ 10^6 Btu/yr
Natural gas ______ 10^3 ft^3	× 1.03	= ______ 10^6 Btu/yr
Oil (no. 2) ______ gal	× 0.14	= ______ 10^6 Btu/yr
Oil (no. 6) ______ gal	× 0.15	= ______ 10^6 Btu/yr
Coal ______ tons	× ______ (~25)	= ______ 10^6 Btu/yr
Other ______	× ______	= ______ 10^6 Btu/yr

Conversion Factors

1 kWh = 3412 Btu
1 ton of cooling = 12,000 Btu/h
1 ton·h of cooling = 12,000 Btu
1 hp = 0.75 kW
1 hp·h = 0.75 kWh

Heat Content

Bunker fuel oil = 150,000 Btu/U.S. gal
Distillate fuel oil = 140,000 Btu/U.S. gal
Natural gas = 1,030 Btu/ft^3
Coal (approximate) = 11,000 to 13,000 Btu/lb

Table 1-4 Metrication

Multiply	By	To Get
Kilowatt hours	3.6	Megajoules
Btu's	1055	Joules
Btu's	1.055	Kilojoules
Btu's	0.2520	Kilocalories
Pounds	0.4535	Kilograms
Cubic feet	0.02832	Cubic meters
Cubic feet	28.32	Liters
Gallons	3.785	Liters
Horsepower	746	Watts

Similarly, 1000 kilowatthours (kWh) of electricity = 3,412,000 Btu (1000 kWh $\times$ 0.003412 $\times$ 10^6, or 3,412 $\times$ 10^6 Btu).

Once the Btu language has been adopted, it becomes possible to make comparisons in the energy performances of buildings. If, for example, we wish to express how one particular office building might perform in two different climates, we can do this graphically, as shown in Fig. 1-2. Or, if we wish to formulate legislation to limit the energy consumption of new buildings by setting up energy budgets, the budget might follow the format of Table 1-5.

The chart shown in Fig. 1-3, from a Department of Housing and Urban Development (HUD) study covering the entire United States, shows the energy consumption (expressed in Btu per square foot per year) for the most efficient building, the most wasteful building, and the average in that category.

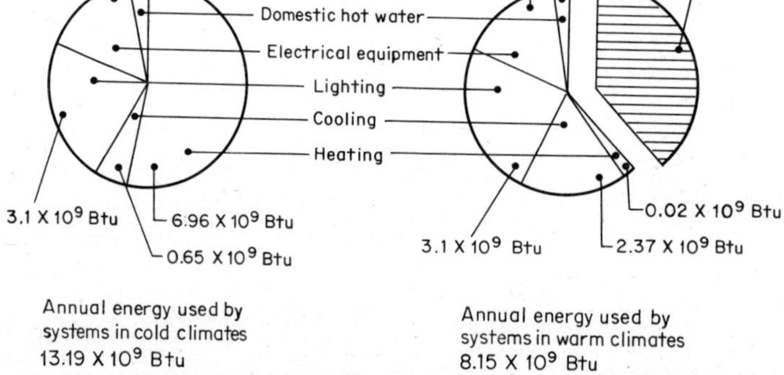

FIG. 1-2 Climate and energy. *(Federal Energy Administration, "Guidelines for Saving Energy in Existing Buildings," 1975.)*

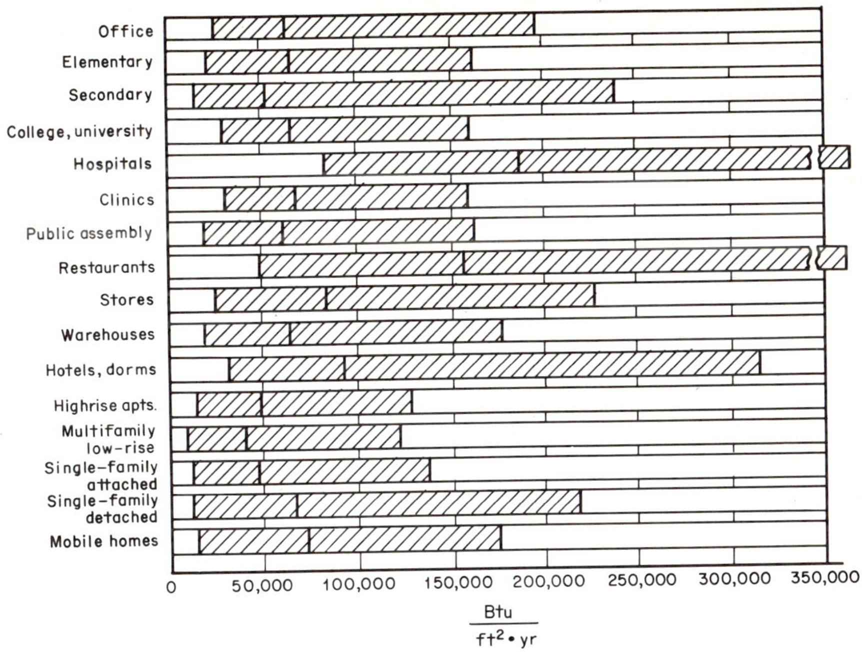

FIG. 1-3 Energy consumption by building types. *(AIA Research Corp./HUD, 1978.)*

Boundary Versus Resource Consumption

The building consumption data previously referred to are known as *boundary consumption*, or the energy supplied directly to the building at its boundary and measured at that point. This can be the energy from a fuel truck delivering oil or from an electric service cable. *Resource consumption*, on the other hand, measures not only the energy delivered at the building boundary, but also the greater energy involved in both originally producing and then delivering the energy to the structure. The average practitioner is concerned primarily with boundary consumption; most government planners, with resource consumption.

In Fig. 1-4 boundary consumption totals 15,910 X 10^6 Btu (9639 + 6271 = 15,910), whereas resource consumption is 28,452 X 10^6 Btu.

Section 12 of ASHRAE Standard 90-75, "Annual Fuel and Energy Resource Determination," provides a methodology for converting boundary into resource consumption.

THE VARIETIES OF EFFICIENCY

The conversion factors in Table 1-3, for determining the number of British thermal units in a given quantity of fuel, list the theoretical values which are obtainable if all the energy contained in the fuel is delivered to the building.

Table 1-5 Maximum allowable energy consumption per year
(Thousands of Btu per gross square foot of conditioned floor area)

UBC Occupancy[1]		Occupant load	Climatic Thermal Zones 1–5			Climatic Thermal Zones 6,8,9,10			Climatic Thermal Zone 7			Climatic Thermal Zones 11, 12, 13			Climatic Thermal Zones 14, 15		
			Heated and cooled	Heated only	Cooled only	Heated and cooled	Heated only	Cooled only	Heated and cooled	Heated only	Cooled only	Heated and cooled	Heated only	Cooled only	Heated and cooled	Heated only	Cooled only
A	Drinking and dining establishments		131	126	82	108	102	82	104	103	80	132	119	96	148	109	126
	All others		159	154	64	123	114	68	118	106	71	155	140	86	189	120	141
	B-1		180	171	163	191	163	184	189	162	184	196	173	185	243	168	236
	Offices	Under 300	141	135	134	145	133	142	145	132	143	148	139	140	145	137	160
		Over 299	126	125	124	129	128	128	131	130	131	134	131	130	132	129	128
	Retail grocery stores		214	212	167	194	189	176	192	187	176	235	216	199	255	204	229
B-2	Other retail stores	Under 300	180	171	163	191	163	184	189	162	184	196	173	185	243	168	236
		Over 299	200	195	190	207	195	202	209	196	204	213	199	205	231	205	255

Drinking and dining establishments		131	126	82	108	102	82	104	103	80	132	119	96	148	109	126
Classrooms		120	118	77	105	94	82	101	94	83	156	143	89	142	97	123
Storage		104	104	104	65	65	65	63	63	63	92	92	92	80	80	80
B-3 B-4 H		104	104	104	65	65	65	63	63	63	92	92	92	80	80	80
E		120	118	77	105	94	82	101	94	83	156	143	89	142	97	123
I[2]		—	—	—	—	—	—	—	—	—	—	—	—	—	—	—
Others[3]		141	135	134	145	133	142	145	132	143	148	139	140	165	137	160

[1]Occupancy, as defined in the Uniform Building Code, 1976 edition. Occupancies A and B-2 have been further subdivided. For B-2 type occupancies not listed, use the subdivision which most closely describes the occupancy.
[2]Maximum allowable energy consumption values are under development for UBC occupancy I.
[3]Occupancies which are not included in any of the listed UBC occupancies.

Source: Proposed Regulations of California Energy Resources Conservation and Development Commission, 1977.

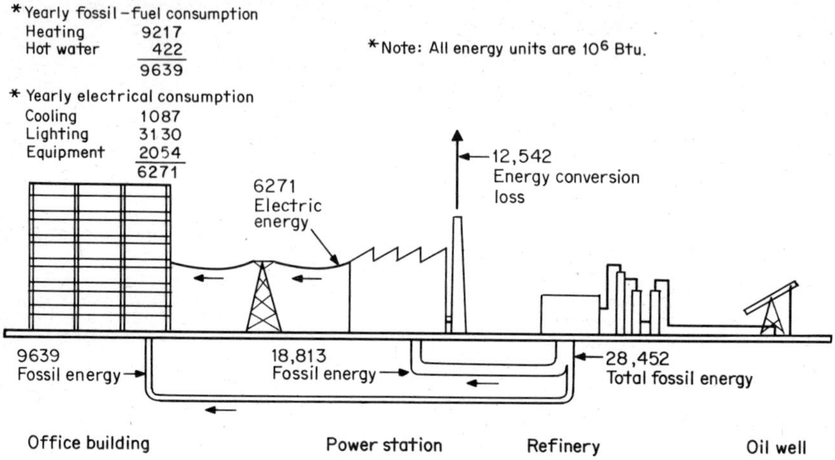

FIG. 1-4 Boundary versus resource consumption. *(General Services Administration "Energy Conservation Design Guidelines for New Office Buildings," 1977.)*

In reality, no system is able to convert *all* the energy. No oil burner burns every last droplet of oil perfectly, and inevitably some heat goes up the chimney. Accordingly, the useful energy actually delivered is always less than 100 percent. Typical efficiencies for properly operating equipment are approximately as shown in Table 1-6. The term steady-state efficiency is used because the efficiency rating is based on the equipment operating continuously at its peak level. In reality, equipment does not operate continuously, but rather starts and stops in response to the demands made on it at different times. In this on again/off again process it loses still more efficiency, in the same way that stop-and-go city driving is less efficient than traveling the same distance on an open highway.

This real efficiency is *seasonal efficiency*, and it reflects climatic and other operating demands that turn systems on or off. In the process, an oil-fired system with a peak steady-state efficiency of 80 to 86 percent may, in fact, have a seasonal efficiency of only 40 to 50 percent.

Steady-state efficiency is used to calculate the boiler sizes needed to meet peak demands, because the boilers will be operating continuously when those peak demands are made on them.

In calculating annual fuel consumption, however, seasonal efficiencies should be used. Data on seasonal efficiencies are generally obtainable in each locality from fuel or energy suppliers. As with all data supplied by sellers of products, it pays to ask hard questions about the basis for the data.

Table 1-6 Equipment efficiencies

Fuel	Steady-State Efficiency (%)
Anthracite coal	65–75
No. 2 oil	80–86
Natural gas	80–84
Electricity (resistance heating)	95–99

Assumptions with respect to seasonal efficiencies are particularly suspect when supplied by competing suppliers and should always be checked in computer printouts which compare the economy of one fuel with another.

FUEL AVAILABILITY AND PRICE

Selection of a particular fuel or energy source must take into account four factors:

1 Availability
2 Usefulness
3 Price
4 Seasonal efficiency

Availability and price vary geographically. In the past these factors have been evaluated in the marketplace in an ongoing process. Thus the decisions made by owners of multifamily units in different parts of the United States are as shown in Table 1-7.

Availability and price will be subject to change as events in the energy world unfold, but all four factors will continue to enter into the selection process.

CHANGING CRITERIA AND STANDARDS

In the days of cheap energy, design criteria used in planning buildings were more concerned with playing it safe than with saving energy. If three air changes could ventilate a space, five might be called for; if 50 footcandles (fc) could produce sufficient light, there was little risked by calling for 100 fc. In effect, we built in a large safety factor which could cushion against shortcomings in design, construction, or operation.

As a result of the energy crisis, we have had to reevaluate past criteria. In doing so—by cutting back—we are starting to operate closer to the bone, and the cushion against shortcomings is becoming thinner.

In view of the proliferation of professional liability suits in recent years, these changes in criteria must be not only discussed clearly with the client, but also reflected accurately in the written minutes of the decision-making process.

ASHRAE STANDARD 90-75 FOR NEW BUILDINGS

The leader in new criteria has been ASHRAE's Standard 90-75, "Energy Conservation in New Building Design." It has been incorporated into the U.S. Energy Conservation Act of 1976. It has served as the basis for model building codes (National Conference of States on Building Codes and Standards, Build-

Table 1-7 Multifamily housing heating fuels in 1970

	Numerical Distribution					Percentage Distribution				
	Electricity	Gas	Oil	Other	Total	Electricity	Gas	Oil	Other	Total
Northeast										
New England	61.4	137.5	297.4	10.2	506.5	12.1	27.1	58.7	2.0	100.0
Middle Atlantic	111.4	869.6	1820.2	135.3	2936.5	3.8	29.6	62.0	4.6	100.0
Total	172.8	907.1	2117.6	145.5	3443.0	5.0	26.3	61.5	4.2	100.0
North Central										
East North Central	143.9	949.2	129.0	238.3	1460.4	9.9	65.0	8.8	16.3	100.0
West North Central	42.0	350.6	34.7	21.0	448.3	9.4	78.2	7.7	4.7	100.0
Total	185.9	1299.8	163.7	259.3	1908.7	9.7	68.1	8.6	13.6	100.0
South										
South Atlantic	337.1	459.8	245.5	69.0	1111.4	30.3	41.4	22.1	6.2	100.0
East South Central	67.9	131.3	2.8	12.9	214.9	31.6	61.1	1.3	6.0	100.0
West South Central	150.4	307.6	1.1	4.4	463.5	32.4	66.4	0.2	0.9	100.0
Total	555.4	898.7	249.4	86.3	1789.8	31.0	50.2	13.9	4.8	100.0
West										
Mountain	45.3	199.9	7.8	8.7	261.7	17.3	76.4	3.0	3.3	100.0
Pacific	348.8	1058.9	69.4	86.3	1563.4	22.3	67.7	4.4	5.5	100.0
Total	394.1	1258.8	77.2	95.0	1825.1	21.6	69.0	4.2	5.2	100.0
Total United States	1308.2	4464.4	2607.9	586.1	8966.6	14.6	49.8	29.1	6.5	100.0

Source: "Incentives for Energy Conservation in Multi-Family Housing," Department of Commerce, 1975.

ing Officials and Code Administrators, Southern Building Code Congress, etc.) and was adopted in varying forms in state and local legislation.

The Standard sets forth requirements for the design of new buildings . . . covering their exterior envelopes and selection of their HVAC, service water heating, electrical distribution and illuminating systems and equipment for effective use of energy.

The Standard covers new buildings that provide facilities or shelter for public assembly, educational, business, mercantile, institutional, warehouse and residential occupancies, as well as those portions for factory and industrial occupancies which are used primarily for human occupancy, such as office space. Unless otherwise stipulated, the term buildings, as used in this Standard, includes mobile homes and manufactured buildings.

Buildings or portions thereof whose peak design rate of energy usage is less than . . . 3.4 Btu/h ft^2 of floor area for all purposes and those which are neither heated nor cooled are excluded from the scope of this Standard.

This Standard does not cover specific procedures for the operation, maintenance, or use of buildings.

Figure 1-5, which sets standards for building envelopes, is typical of those contained in ASHRAE 90-75.

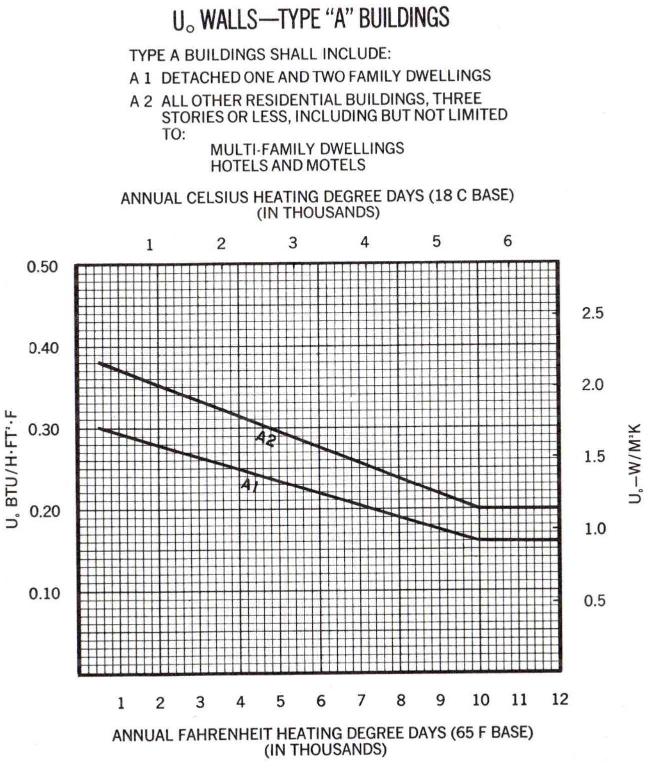

FIG. 1-5 Thermal transmission of type A building walls. *(Reprinted with permission from ASHRAE Standard 90-75, Energy Conservation in New Building Design.)*

ASHRAE STANDARDS FOR EXISTING BUILDINGS

The purpose of these Standards, called the 100 Series, is to provide requirements and criteria for the conservation of energy in existing buildings.

Standard 100.1P: Low-Rise Residential

This Standard is applicable to all existing detached single-family and two-family dwellings including those with not more than five (5) lodgers or boarders outside the family unit per family. Included also are mobile homes, all tri-plexes, four-plexes, row-houses and other multiple family dwellings not more than three (3) stories in height exclusive of basements and having complete kitchen and bath facilities for each dwelling unit.

Standard 100.2P: High-Rise Residential

This Standard is applicable to all existing structures used as high rise residential buildings. Included are all multiple family dwellings having more than three (3) stories, exclusive of basements, and having complete kitchen and bath facilities for each dwelling unit, and having no more than five (5) lodgers or boarders outside the family unit per family. Included, among others, are apartment buildings, condominiums, medium rise garden apartments, etc., where non-transient occupancy exists. Included also are all buildings that can be reasonably interpreted as being of this type, even if they are not specifically named in this Standard.

Standard 100.3P: Commercial

This Standard is applicable to all existing structures used as hotel and motel buildings, lodging houses, boarding houses and dormitory buildings arranged for the shelter and sleeping accommodations of one (1) or more individuals excepting those classified as Institutional, Low Rise Residential or High Rise Residential. Complete kitchen and bath facilities may exist if the lodging is for vacational or other transient purposes as compared to residential purposes. Also included are all structures used for display and sales purposes involving stocks of goods, wares or merchandise incidental to such purposes and accessible to the public, including, among others, retail stores, shops, salesrooms and markets. It shall also include all structures used for the transaction of business, for rendering of professional services or for other services that include stocks of goods, wares or merchandise in limited quantities incidental to office use or sample purposes including, among others, offices, banks, civil administration activities (such as post offices and courthouses), radio and TV studios and stations, motor fuel service stations and similar establishments. Also included are assembly halls, restaurants, fast food concessions and cocktail and snack bars designed to contain less than fifty (50) persons at capacity. Included also are all buildings that can be reasonably interpreted as being of this type, even if they are not specifically named in this Standard.

Standard 100.4P: Industrial

This standard is applicable to all existing structures used as industrial buildings. Industrial buildings shall include all structures in which occupants and/or machines are engaged in fabricating, assembling or processing of products or materials, including, among others, factories, assembly plants, industrial laboratories and all others used for industrial and manufacturing purposes. Also included are structures designed to house, among others, the production and distribution of electric, gas or steam power, telephone exchanges, rolling mills and foundries. Also included are structures for the storage and warehousing of all materials and products not usually available for retail sale and requiring energy for environmental control. Inculded also are all buildings that can be reasonably interpreted as being of this type, even if they are not specifically named in this Standard.

Standard 100.5P: Institutional

This Standard is applicable to all existing structures used as church schools, schools, colleges, and for similar educational purposes. Also included are all research institutes, and the testing and research laboratories incidental to their use and for educational purposes. Also included are all structures harboring people suffering limitations because of health or age for care or treatment, including, among others, day nurseries, hospitals, sanitariums, clinics, infirmaries, orphanages, and homes for the aged and infirm. Also included are structures in which people are detained for penal or correctional purposes or in which the liberty of the occupants is restricted, including, among others, jails, prisons, reformatories, and other mental health institutions. Also included are structures used for prosecuting public or civic services and activities of an emergency nature, including, among others, fire houses and police stations. Also included are museums, galleries, libraries and similar buildings. Included also are all buildings that can be reasonably interpreted as being of this type, even if they are not specifically named in this Standard.

IMPACT OF ASHRAE STANDARDS

In a study prepared by Arthur D. Little, Inc., for the Federal Energy Administration (FEA), a comparison was made between buildings built prior to the oil embargo of 1973–1974 and the same buildings modified to conform to the new ASHRAE Standards.

The conclusions reached were as follows.

1. Impact on Building Energy Consumption

Under a strict interpretation of the standard, ASHRAE 90 would reduce annual energy consumption in all building types and locations, particularly in nonresidential structures. In comparison with 1973 construction and operational practices across the four locations investigated, annual energy consumption (unweighted averages) would be reduced as follows:

Single-family residence	-11%
Low-rise apartment building	-43%
Office building	-60%
Retail store	-40%
School building	-48%

The apparent lower effectiveness of ASHRAE 90 for the detached single-family residence may be explained in part by the moderately high overall thermal efficiency assumed for the conventional residences.

2. Impact on Building Characteristics

The application of the standards brought about the following physical differences in the conventional versus modified buildings:

Exterior Glass—Under ASHRAE 90, glass area (percent fenestration) was reduced in approximately two-thirds of the buildings. Reductions were as much as 30%, but most were less than 20%. One region—the North Central— required reductions in glass area for all buildings. Double-glazing was required in virtually all Northeast and North Central construction.

Exterior Wall—Decreases in glass area under ASHRAE 90 were balanced by increases in net wall area; virtually all increases were less than 8%.

Insulation—Additional insulation requirements for residential construction varied from less than 80 to as much as 300 pounds per unit for both standards. Increased requirements for insulation in commercial construction due to ASHRAE 90 were even greater than those needed in residential construction.

Lighting—Reductions in lamps and lighting fixtures were limited to ASHRAE 90 and varied by building type, averaging 24% and 22%, respectively, for nonresidential construction.

HVAC Systems Capacities—Under ASHRAE 90 reductions in heating system capacities averaged 42% for the five building types, while reductions in cooling systems averaged 31%. The greatest reductions were found in school buildings. The rated kilowatt capacity of auxiliary HVAC equipment—pumps, towers, fans, supply fans, etc.—were reduced an average of 44%.

Chapter 2

Incorporating Conservation into the Building Process

Buildings differ from one another physically, are erected in many locations, and are built at various times. Despite these variables, the design/build process by which structures are created follows a fairly constant pattern. This similarity makes it possible to outline ways in which conservation techniques can be incorporated into the building process throughout a variety of situations.

THE BUILDING PROCESS

In its simplest form, the process has three participants: a client with a set of needs and the money to satisfy them, a designer to transform these needs into a set of building instructions, and a builder to erect the structure in accordance with the instructions.

Successful conservation techniques have three aims: to make the client both aware of conservation as one of many needs and willing to provide the money to satisfy it; to create a similar awareness in the designer to include conservation as a design consideration; and to ensure that the builder capably incorporates these determinations into the final structure.

The process starts well before a designer enters the picture; it begins with a client in need.

It is an axiom of the building process that the earliest decisions are the most important; they set the basic directions and parameters within which all that follows must operate. Yet these early decisions involve much smaller amounts of money and workforce than the later activities of formal planning and construction. Rarely does a feasibility study cost as much as a set of plans or the final structure.

A client stating needs establishes a scope of work which sets the outside limits of energy consumption. When a selection is made of a geographic region or a particular site on which to build, it not only determines the climatic environment, but also subjects the client to the availability of local fuels and the rate schedule to be followed in paying for energy purchased from local utility companies.

Despite their importance, these early decisions are the easiest to alter, since commitments of time and money are still at a minimum. But as the design/build process continues and increasing expenditures are made, a momentum builds up which makes change increasingly difficult.

It is helpful, therefore, to begin the description of the process by examining the characteristics of the various participants in decision making at different stages in order that their functions can be understood and their assistance at moments of change facilitated.

The Participants

The number of participants varies with the complexity of the project. A do-it-yourself homeowner is client, designer, and builder all in one. The creation of a new town involves thousands of people. The following paragraphs describe functions which may be embodied in any number of individuals in any particular case.

The Client

The client is the boss, the one whose needs originate the project and whose money ultimately controls all decisions. The boss builds because a home is needed for a family, a factory for its products, or classrooms for their students. This individual may care little about energy considerations except under the threat of fuel rationing or cutoff or when escalating energy costs impact on the operational budget. However, these events are increasingly likely. The boss is receptive to government tax or other financial incentives aimed at encouraging energy conservation.

Client Staff (Operations and Maintenance)

Traditionally the concern of client staff has been primarily the smooth functioning of the structure; energy expenditures have been only a secondary consideration. The energy crisis is changing this emphasis, and increasing numbers of staff are assuming the initiative in energy conservation.

Architect

The architect is chief of the design team. He or she transforms the client's needs into the building concept initially and then into the construction documents from which the structure is built. The architect's design concept has basic energy consequences and establishes the limits within which all subsequent energy decisions are made.

Mechanical Engineer

The mechanical engineer is usually hired by the architect as part of the design team. She or he has responsibility for plumbing, heating, ventilating, air conditioning, electrical, and other mechanical systems required by the building or its occupants. The actions of the client and architect create the energy demands which the mechanical engineer satisfies. This consultant's early input into client-architect decision making is most desirable.

Energy Consultant

An emerging speciaist, now small in number, created by the energy crisis is the energy consultant. The job of the energy consultant is similar in function to that of construction manager but limited to energy considerations. Functions include programming, planning, technology, construction supervision, financing, and operations.

The term *energy auditor* is increasingly used by government agencies to describe energy professionals who carry out studies mandated by various regulations. In setting up its rules for making energy audits, the FEA states that the auditors either must be state-licensed professional engineers or architects or must have an engineering or architectural degree from an accredited college or university and four years subsequent experience in the design and/or operation of buildings.

Utility/Fuel Suppliers

Utility/fuel suppliers are an excellent source of data on the future supply and probable cost of their products. Their comparisons of their own product versus their competitors' should be accorded the same consideration given sellers of any product.

Banks

While legislation which will provide government guarantees for energy-related loans (see Chap. 15) may change future thinking, a bank relies on the basic financial strength of the borrower's ability to repay a loan and tends not to be overly influenced by energy considerations. Despite this fact, energy-conscious expertise displayed by a borrower often helps create a favorable impression.

Government Agencies

This group includes those agencies with direct energy responsibility [the Department of Energy (DOE), FEA]; those whose activities affect energy consumption in building types controlled by them [HUD; the Department of Health,

Education, and Welfare (HEW); and the General Services Administration (GSA)]; and general regulatory agencies whose codes and regulations establish criteria with significance to energy consumption [building codes, National Board of Fire Underwriters (NBFU) regulations, professional and trade association procedures].

In emphasizing the importance of early decisions on the ultimate consumption of energy, it would be naive to expect energy considerations to be the overriding factor in the major decisions of what a building is going to be like. No one ever started out to build a building in order to save energy.

It is reasonable to expect, however, that the energy consequences of such early decisions as site selection and basic building concept be called to the attention of the participants in the process so that they are aware of the consequences of their actions. In borderline decisions or where more than one alternative makes sense, the energy consideration can sometimes tip the balance.

Among the most important early actions undertaken by the client is an examination of expected operations with respect to intensity of use. In our affluent culture we have long enjoyed the luxury of specialized buildings with specialized spaces which are used at intermittent periods but which require energy round the clock.

In setting up the program for the building, the client should consider intensity of use with respect to both space and time.

Space

The multipurpose room used intensively is always more efficient than several spaces used separately. This efficiency applies to construction and maintenance costs as well as energy consumption.

Time

There are 168 hours (h) in a week. A 40-h work week thus uses a structure less than one-quarter of the time it is available. To the extent that this percentage of occupancy can be increased, the size of the structure can be reduced, though not necessarily in direct proportion. Buildings containing large computer installations already run round the clock to help amortize computer cost; retail stores are beginning to run on similar schedules for competitive marketing reasons.

Additionally, the use of buildings round the clock allows freer access to off-peak utility rates, which are increasingly being set lower than the charges for the same energy during traditional work hours. These lower off-peak rates reflect the ability of a utility to use its physical plant during hours in which it would otherwise be idle.

PROFESSIONAL LIABILITY

One aspect of energy conservation that may have consequences in the area of professional liability and malpractice actions derives from the innovative nature of the techniques and equipment being introduced. In the event of failure, new techniques invite legal assault more readily than traditional procedures even though those new techniques were not, in fact, the cause of the problem.

Accordingly the usual caveats with respect to written records of the decision-making process should be particularly enforced with respect to energy decisions.

Design

Selecting the Team

The energy core of the design team consists of the architect, the consulting mechanical engineer, and the energy consultant.

The team should be set up as early as possible. When the client has a continuing relationship with these professionals, their input can begin as a result of a phone call or a conversation at lunch. When no such relationship exists, many clients are reluctant to become involved with an architect until there is a definite job to offer. By holding off, they deny themselves advice at the time it is most needed. Most professionals welcome the chance to be consulted at these early stages before a job has crystallized, since it gives them an inside track should the job become reality. Many view this early consultation as part of public relations promotional activity for their firm.

Traditionally the architect has retained the mechanical engineering consultant. No set procedure as yet applies to the energy specialist. Some clients will prefer to do this hiring directly, particularly where there are energy concerns of great magnitude. By doing so they ensure that the design team will have the benefit of this input, but because the energy specialist is reporting directly to the client, the result is a continuing goad and critic throughout the design/build process. Regardless of who does the retaining, the energy specialist must be acceptable to and compatible with the balance of the team.

It should be noted that most architectural and engineering firms operating today were established at a time when energy conservation played a small part in design deliberations. Some of these firms are making strong efforts to alter their office procedures to reflect these changing times; others are not. It is incumbent upon the client in selecting architects, and upon the architect in selecting mechanical engineers, to probe their recent activities in this area.

The Design Sequence

The American Institute of Architects (AIA) Agreement between Owner and Architect divides design activity into five phases:

Schematic design
Design development
Construction documents
Bidding or negotiations
Construction administration

SCHEMATIC DESIGN Schematics is a time for becoming acquainted, for examining requirements, for exploring assets, for investigating controls and setting timetables—in short, for charting course and setting sail.

Energy conservation in the finished structure will benefit greatly if at the kickoff meeting the client (everyone's boss) makes a simple statement to the effect that energy conservation is a matter of concern and asks the team members for their off-the-cuff thoughts on the subject. Whatever the depth of the ensuing discussion, its introduction at the kickoff meeting will plant its importance firmly in the minds and subsequent activities of the designers throughout the project.

Clients should also be prepared to state their position with respect to life-cycle costing (see Chap. 3) and the extent to which they are prepared to make a higher initial investment in order to gain future saving. Clarity in presenting this attitude will facilitate the consideration of financially realistic options in energy design. Finally, the kickoff meeting should designate responsibility for the monitoring of energy decisions throughout the development of the job. Normally this will be the function of the energy consultant, but should there be none on the team, the responsibility will fall to either the architect or the mechanical engineer. In this case it must be made clear that the professional designated is responsible not only for the energy consequences of his or her own work, but also for seeing that the other members of the team live up to their obligations.

Once the energy leader has been designated, the following information should be gathered and made available for discussion.

- Current and projected availability of energy fuels (oil, gas, electricity) and similar data with respect to energy prices or rate schedules. Data should be obtained from at least two independent sources for purposes of verification.
- Current energy consumption by buildings similar in scope. These data will provide energy budgets that can be valuable points of reference as the design process moves along.
- Current energy conservation techniques being employed in similar structures. In a fast-changing field like energy, constant updating is critical; and the latest input from professional, government, trade, and academic sources should be sought.
- Recent changes in design criteria. Codes and criteria which affect energy are undergoing substantial revision in the wake of the energy crisis. These constraints set most of the standards under which mechanical equipment is currently designed. They involve both specifications and performance. Because they generally require input from many affected parties, code revisions come into being slowly. Where, as in most energy revisions, the changes are being introduced to permit less energy consumption, a judgment should be made as to which rules are likely to be in effect at the completion of the structure, and the design should reflect these.
- Implications of alternate energy sources (solar, wind) with respect to the proposed structure. If any feasibility is indicated, then space, structural, and cost parameters should be prepared.

Once these data have been discussed, their implications for the design of the structure should be analyzed and assessed. The architect, whose responsibility extends to all aspects of the building design and not just its energy aspect, may not be able to accept all energy-related recommendations; but the options should be made available.

To the architect, these energy implications are but one of a host of considerations that enter into the design process as the first-concept sketches are prepared. Until the past century, when the introduction of electricity and fossil fuels gave us the energy to power new mechanical equipment, architects traditionally plannod buildings to minimize the impact of the environment and designed their structures to be responsive to it. The characteristics of regional architecture throughout the world reflect these climatic considerations, characterized as "passive design," and were long a part of architectural education and practice. Passive design considerations include:

- *Siting and orientation:* The relation of the building to the land, the sun, and the wind.
- *Building shape:* The less skin, the less exposure.
- *Nature of the envelope:* Fenestration, insulation, thermal mass, wall shading, color and reflectivity, openings and penetrations.

Just as the introduction of energy fuels gave us the ability to ignore many of the constraints of passive design and devote attention to other considerations, so the energy crisis is forcing us to return to our traditions. Architectural schools are examining their curricula in order to emphasize again the lessons learned over many centuries of scant energy, and practicing professionals are similarly looking backward. The importance of returning to these passive fundamentals cannot be overemphasized.

Figure 2-1 illustrates the changes that result from applying principles of passive design. The Federal Office Building in Manchester, New Hampshire, was designed by Isaak and Isaak, Architects with Dubin-Mindell-Bloome Associates as energy conservation consultants. Figure 2-1 shows the original "conventional" design and the final "energy conservation" building. A fuller discussion of these changes appears in Chap. 13.

The concept produced by the end of the schematic phase should, therefore, reflect the principles of passive design together with implications of the energy update data provided by the energy leader of the design team.

DESIGN DEVELOPMENT If the schematic phase is a time for broad-brush beginnings, design development starts getting down to the nitty-gritty. This is the time when various energy options are explored and assessed, and final strategies agreed upon. By the end of this phase the die has been cast, and substantive change from this point forward is possible only with trauma.

Conclusions are reached only after all possible options have been explored. Yet because the work of each of the design professionals affects that of the others so strongly, a kind of cat-and-mouse game tends to develop. If the architect is

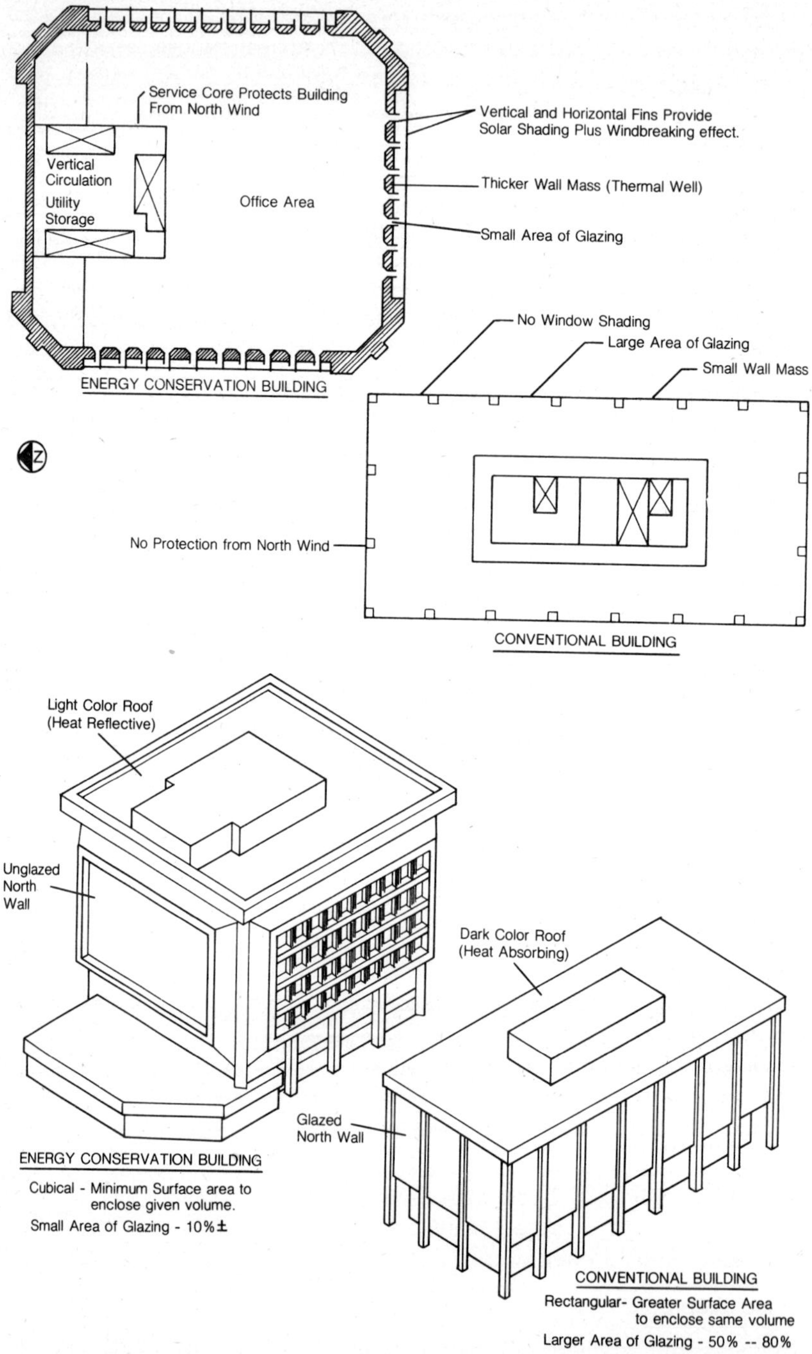

FIG. 2-1 GSA office building in Manchester, New Hampshire. *(General Services Administration, "A Case Study—Designing an Energy-Efficient Building," 1975.)*

going to be asked to provide a south-facing sloped roof to support a battery of solar collectors, it would be helpful to know this before she or he makes a false beginning by starting to design a building with a flat roof. Similarly, the mechanical engineer would like to have a pretty good idea of spatial shapes and aesthetics before beginning the design of lighting systems.

One of the dangers of this phase, particularly as it affects energy, occurs when the architect pursues and completes design development work with only minimal input from the other consultants; then, having "fixed" the architectural design, the architect turns it over to the other members of the team for their input. Most mechanical engineers can make almost any architectural scheme work, but if the structure has been fixed in most of its details before they begin, their freedom to maneuver has been cut out from under them, and the final energy efficiency of the structure is bound to suffer.

It is far better to reverse this traditional sequence. Most schematics are in sufficient detail to permit the consultants to satisfactorily explore the options, and any items needing amplification can be made available by the architect. With this procedure, the consultants are free to consider the full gamut of options. Having reached their conclusions, the consultants present and discuss them with the architect for incorporation into the work.

This change in sequencing is suggested because much of the hardware involved in energy conservation, such as heat wheels or thermal storage devices, is new to many architects. Thus they are not yet familiar with the space, structure, and location required by this equipment. Moreover, this uncertainty is bound to continue since the vast amounts of energy research currently being conducted will inevitably produce a continuing flow of new products and technologies.

Having conducted their investigations first, the consultants will be able to advise the architect on such matters as:

- *Fenestration:* The percentage of openings for each orientation.
- *Insulation:* For walls, roof, and cellar levels, at interior partitions separating spaces which are treated differently.
- *Thermal mass:* Desired densities for wall and interior structure.
- *Space and structural requirements*: Together with optimal locations, for major items of equipment such as reservoirs for energy storage.
- *Horizontal space requirements:* Ceiling and roof plenums.
- *Vertical space requirements:* For ducts, shafts, and insulation.
- *Exterior equipment:* On roofs, adjacent to building on the ground, louver openings, and any other items visible on the outside of the building.
- *Reserve spaces:* Not needed for the energy supply situation projected for initial building occupancy, but desirable to accommodate future changes deemed probable in the light of continuing changes in energy availability.

Given this thinking early in the design development phase, the architectural work can be carried on with energy consequences being given full consideration.

When this phase has been completed, it would be prudent to check its thinking by requesting from the consultants a projected energy budget for the structure and its systems as fixed in the design development documents. Further, this material should be, wherever possible,

- Compared with the energy budgets for similar structures previously assembled during schematics.
- Submitted to relevant government agencies and to fuel suppliers for comment.
- Submitted to firms likely to be retained for equipment maintenance of the completed structure.
- Priced with respect to the added cost of energy-conserving equipment so that its inclusion can be verified with the client, who will then be better protected against cutbacks during the construction bidding and negotiation stages.

CONSTRUCTION DOCUMENTS During this phase every element of the structure is committed to paper. This activity is carried out by large numbers of staff, many of whom may not have participated in the broad exploratory investigations of the earlier decision-making phases. While it is unlikely that this phase of the work will introduce any dramatic improvements, it can very often inadvertently negate much previous accomplishment. Thus, a mechanical drafter working on sizing ventilation ducts may reach for a handbook containing duct and fan sizes based on criteria which the top-level discussions held during schematics had decided was outmoded. Or an architectural detailer, not realizing that insulation is to be applied on the exterior of walls (as is beginning to be done in certain thermal designs) will place it on the inside surface as was always done previously.

Eternal vigilance, always appropriate, is particularly important during this stage, and the leaders in drafting room production must be reminded to review the minutes of the previous phases so that their value and effort are not eroded.

BIDDING OR NEGOTIATIONS A normal part of the bidding process is the receipt of phone calls from contractors seeking to clarify certain aspects of the construction documents. Since energy technology is in a continuing state of flux, its equipment and procedures are constantly changing, and questions relating to this work will be particularly numerous during this period. Moreover, since the traditional bidding process emphasizes lower initial costs rather than lower life-cycle costs, the calls may be greater than usual.

Since budgets are always tight, a list of alternatives which has been well thought out should be included as part of the bid package. This is much preferred to the receipt of suggested cuts from contractors who, new to the job, cannot be expected to be aware of its long-range aims.

CONSTRUCTION ADMINISTRATION Construction begins when the shovel hits the ground. It ends when the client moves in.

Construction of the building by the contractor, and the supervision of construction by the design team, is basically no different for an energy-efficient building than for any other kind.

A few energy-related suggestions can be made for this phase.

Mechanical equipment is often late in arriving at the job. Many energy-related installations rely on the efficient operation of groups of equipment, and not only on isolated pieces. It is essential that the desired potential be achieved by adequate adjustment and tuning up of the entire installation; one should not fall victim to the last-minute rush to move in.

It is advisable for the staff who will operate the finished structure to be on hand while the final tuning up is being accomplished. They should be given the appropriate equipment manuals and operating instructions; and they should be acquainted by the design team with the thinking that went into the design of the structure and the way in which it is to be operated.

The use of new diagnostic tools to check the tightness of the envelope is desirable. A thermographic scan of the enclosed structure will often reveal leaks and gaps in insulation and caulking not noticed by an inspector. (Examples of thermography are given in Chap. 10.)

OCCUPANCY AND OPERATIONS

Initial occupancy of a structure is like the shakedown cruise of a ship. The responsible client will see that the operations staff keeps in touch with the design team and contractor during the start-up year (which coincides with the usual warranty period and includes both summer and winter operations).

It is advisable to schedule one or two meetings of all these parties during this first year even if no problems develop, in order to alleviate the plight of a maintenance superintendent who suffers in silence or who modifies the system to act in ways it was never intended to operate.

After a year has passed, copies of the energy bills should be sent to the design team for comparison with design projections, for accuracy of billing and for suggestions on improving performance.

Finally, since at least three years will have elapsed since the original energy-related decisions were made (one year each for design, construction, and initial occupancy), the designers should be asked for a memorandum on advances in energy thinking since the original conception.

Chapter 3

Evaluating Conservation Options

A properly designed, energy-efficient building will have a lower initial cost than one planned disregarding energy consequences. This cost advantage derives from:

- Smaller building volume resulting from more intensive use of space.
- Lower energy demand as a result of incorporating principles of passive design into the basic architecture.
- Design of mechanical systems in accordance with energy-efficient criteria.

By and large, the construction industry operates on the basis that low initial cost is the only way to survive in the competitive marketplace. Exceptions to this generalization range from the institutional client who is seeking image rather than utility to the well-capitalized and sophisticated corporation that can afford to invest for long-range benefit. These exceptions to the contrary, the pressures to reduce initial cost govern most marketplace action.

Nonetheless, the energy crisis is modifying this attitude by forcing us to consider the consequences of:

- Fuel shortages and/or rationing. No fuel is cheap, however low its price, if it's not available.
- Energy-saving construction and equipment devices with such short payback periods (less than three years) that their initial cost is so rapidly offset

by their energy savings that they merit consideration as additions to the basic construction budget.

- Tax credit and investment incentives being enacted by government in response to the energy problem which, in effect, provide the money for energy-saving expenditures with longer payback periods.

FUEL COMPARISONS

The operating cost of a fuel is a function of its purchase price, the amount of energy contained in it, and the efficiency with which a mechanical plant can extract and deliver that energy to the building.

The purchase price of fuel is a function of the marketplace, both foreign and domestic. The marketplace operates against a background of political as well as economic considerations.

The amount of energy contained in a fuel is expressed in British thermal units. One therm is 100,000 Btu. For the heat content of various fuels see Table 1-3.

The efficiency with which a mechanical plant can extract energy from a fuel is a function of the design and operation of specific plants. Data regarding efficiencies should be carefully verified to confirm that the conditions assumed in establishing efficiency ratings (combustion air supply, insulation, rate of operations, etc.) are actually satisfied by the structure.

Operating cost is not the only criterion for selecting a heating fuel. Capital cost of the plant, plant maintenance, fuel availability, fuel storage capacity, and environmental impact must be weighed along with fuel cost per therm.

EVALUATING OPTIONS WITH SHORT PAYBACK PERIODS

Conservation options with short payback periods (less than three years) are more likely to be adopted than those whose benefits must accumulate over many years. Short payback options produce higher rates of return on investment, faster return of capital for reuse, and comparative freedom from the hazards of error implicit in the long-range forecasting required by extended life-cycle benefit analysis.

All these factors result in greater credibility, and therefore acceptance, for the proposal which offers fast results. Capital is needed for many purposes; competition for its use is always intense. And if conservation options are to be adopted, their credibility must be strong. Typical conservation options are discussed in Chaps. 5 to 10.

To calculate the payback period for a proposed option, first determine the first cost (FC) of the proposed option, the annual fuel savings (AFS), the projected fuel price (PFP) over the life of the project, and the annual operating cost (AOC) of the proposed option. Next calculate:

$$\text{Annual net savings (S)} = \text{annual fuel savings} \times \text{projected fuel price} - \text{annual operating cost}$$

or

$$S = AFS \times PFP - AOC$$

Then

$$\text{Payback period (PP)} = \frac{\text{first cost (FC)}}{\text{annual net savings (S)}}$$

Example 3-1 Calculating the Payback Period for a Proposed Option

Given the following facts:

Item	Standard Insulation	Proposed Insulation	Difference
Insulation cost	$1,931	$ 6,817	
Heating equipment cost	7,600	3,650	
Total	9,531	10,467	$936
Annual fuel cost	4,964	2,375	−$2,589

From the above: FC = $936

S = 2589 (There is no AOC as a result of additional insulation)

$$PP = \frac{FC}{S} = \frac{936}{2589} = 0.36 \text{ yr (or 4 months)}$$

EVALUATING OPTIONS WITH LONG PAYBACK PERIODS

As a matter of national energy policy, increasing legislation is coming into existence which encourages, primarily through tax credits and investment incentives, the funding of energy options whose payback periods exceed the three years discussed in the preceding section. The development of renewable energy sources (solar and wind among them) currently requires many times this short period to recoup the initial investment.

Over these longer periods comparison on the basis of a simple payback period (PP) is subject to criticism on several counts. Comparisons between alternative investment opportunities of different lifetimes cannot be made; and dollars saved in future years are credited the same as dollars saved in current years, thus ignoring inflation.

Return on investment (ROI) is a somewhat superior procedure because it takes into account the depletion of the investment over its economic life by providing for renewal through a depreciation charge. Using a straight-line depreciation charge (DC) where

$$DC = \frac{\text{first cost (FC)}}{\text{estimated lifetime (EL)}}$$

the percent return on investment can be calculated using

$$\text{ROI (\%/yr)} = \frac{S - DC}{FC}\,100\%$$

The return on investment has the advantage of putting investments with different life expectancies on a comparable basis. It is frequently used in the financial analysis of potential investments because of its simplicity of calculation.

Example 3-2 Calculating the Return on Investment for a Proposed Option

Management is considering a capital investment in its manufacturing process for energy conservation purposes which will cost \$100,000 to design and install but will involve no new recurring costs. This project is expected to save an average of 27,500 MBtu/yr of natural gas for the next 10 yr. The projected average cost of this fuel during the period is assumed to be \$1.00 per MBtu. Assuming that management feels that a 20 percent discount rate is appropriate, will this be a profitable investment?

$$\begin{aligned} FC &= \$100{,}000 \\ AFS &= 27{,}500 \text{ MBtu/yr} \\ PFP &= \$1.00/\text{MBtu} \\ S &= AFS \times PFP - AOC \\ &= \left(27{,}500 \frac{\text{MBtu}}{\text{yr}}\right)\left(\frac{\$1.00}{\text{MBtu}}\right) - 0 \\ &= \frac{\$27{,}500}{\text{yr}} \end{aligned}$$

The payback period (no discounting) is

$$PP = \frac{FC}{S} = \frac{\$100{,}000}{\$27{,}500/\text{yr}} = 3.6 \text{ yr}$$

The return on investment is

$$DC = \frac{FC}{EL} = \frac{\$100{,}000}{10\ yr} = \frac{\$10{,}000}{yr}$$

$$ROI\left(\frac{\%}{yr}\right) = \frac{S - DC}{FC}\,100\%$$
$$= \frac{\$27{,}500/yr - \$10{,}000/yr}{\$100{,}000}\,100\%$$
$$= \frac{17.5\%}{yr}$$

Using return on investment as an approximation of the profitability of this project, we see that even after an allowance for depreciation this appears to be an attractive investment. Additional or second-level measures of performance are needed, however, if we wish to incorporate the time value of money into the analysis.

Second-level measures of performance are those which incorporate an allowance for the time value of money, generally in the form of a discount factor. Because of alternative investment opportunities, a dollar held today is worth more than a dollar held in some future time period. The internal rate of return on the best available investment alternative is generally considered to be the appropriate discount rate for evaluating new investment opportunities, unless this rate is below the true borrowing rate when a new investment needs to be financed. In this case, the discount rate must be at least as high as the borrowing rate.

While appropriate discount rates may differ widely in different industries and even among firms within the same industry, corporate discount rates usually run from 10 to 20 percent or higher. Again, this is equivalent to saying that such a return can be realized elsewhere, and thus for a new investment to be justified it must yield a return somewhat greater than this.

Several second-level measurements for evaluating options are available. These include benefit/cost analysis, time to recoup capital investment, and internal rate of return.

Benefit/cost analysis requires the direct comparison of the present-value benefits (savings) generated by a given investment with its costs. Generally this is formulated in terms of a benefit/cost ratio (B/C). A ratio greater than unity implies that the expected net benefits (properly discounted and summed over the lifetime of the investment) will exceed the initial costs, and therefore such an investment is profitable. Likewise, a benefit/cost ratio less than unity implies that such an investment is not profitable. As an absolute measure of the profitability of an investment, this is generally considered superior to all others.

Time to recoup capital investment, or the "break-even" period, is similar in concept to the payback period (PP) discussed earlier, except that the break-even period takes discount rates into consideration. Again, the chief disadvantage of such a measurement is that investments of unequal lifetimes cannot be compared. However, this measurement of performance is often useful to financial planners and budget analysts.

The internal rate of return (IRR) is defined as that discount rate which reduces the stream of net returns associated with the investment to a present value of zero. While in general the IRR is not always a good measurement of economic performance, it will give good results in evaluating a project which has a fixed first cost followed by a stream of positive net benefits.

Most of these second-level procedures are invoked in order to investigate—and sometimes to justify—energy conservation options which, having long payback periods, require these fairly sophisticated accounting techniques in order to prove their advantage. In the business marketplace the ability of these second-level procedures to divert corporate funds for energy purposes is minimal, if the case could not otherwise be made on the basis of the much simpler payback period or return on investment comparisons (they are, of course, helpful in amplifying the financial implications of proposals which have already gained acceptance).

Where still greater cost-benefit detail is called for, a life-cycle cost study can be undertaken. This is an evaluation which takes into account costs over the specific period known as the "life cycle." Standard accounting texts are available for reference.

Regardless of the method chosen, the analyst should always make clearly known the assumptions made about each of the relevant factors. Output is only as reliable as input.

ENERGY BUDGETS

The preceding sections have dealt with ways of evaluating different energy conservation options. The options considered generally involve component parts of a structure, whether they be in the category of general construction (insulation, glazing, etc.) or mechanical equipment systems and subsystems.

Only through this comparative process can the designer know that all the possibilities have been considered and judged equitably.

But what about the bottom line?

The total energy consumption of a building is the sum of all the energy consumed by its component parts. This total, the energy budget for the entire building, is really everyone's basic concern. The components whose summation produce the total are necessary steps along a path, but not the final goal.

In comparing the energy budget for one's own building, be it new or a retrofit, to the energy consumed by comparable existing structures, one is quickly confronted with variables that, on the face of it, would appear to make direct comparison extremely difficult.

Geographic location with its climatic variation, differences in types of building occupancy (home, school, stores), and hours of use (daily, weekly, seasonally) are just some of the many variables that seem to make bottom-line comparisons difficult.

On closer examination, however, most of these variables will disappear. If a school is proposed for construction, a summary of the energy budgets for a number of neighboring schools can be revealing.

Energy budgets are usually converted to British thermal units per square foot per year, for purposes of comparison. Where available data are expressed in terms of fuel, i.e., gallons of oil or kilowatthours of electricity, these can be converted to British thermal units by Table 1-3.

After the data are compiled, a visit to one or two of the schools with the lowest energy budgets will be revealing. Because they are local there are no climatic differences. Because they are schools, their general occupancy is similar.

A number of variables still remain, including hours of use, building configuration and type of construction, specifics of mechanical and electric systems including off-hour procedures during nonoccupancy periods.

But it is precisely these remaining variables that probably account for the differences in energy consumption. Therefore, it is particularly appropriate that they form the basis against which the proposed structure is to be measured.

Data sources for energy budgets are expanding rapidly. These include government energy agencies, management firms and trade associations, fuel and utility companies.

One caution should be observed: In compiling energy budgets for existing structures, do not rely on preoccupancy design calculations. Get the data based on actual consumption.

In addition to comparisons with the energy budgets of existing structures, the concept is emerging of legislated energy budgets within which proposed structures (and, in time, even existing ones) will have to operate. For example, GSA has set a goal of 55,000 Btu/(ft^2 · yr) for its new office buildings. HUD, in its minimum property standards for one- and two-family dwellings, limits heat loss to 20 Btu/(ft^2 · h).

Legislation already on the books, such as the Energy Conservation Act of 1976 as well as measures by many of the states, is leading to the promulgation of budgets for many areas of construction.

Proponents of budgets argue that they accomplish the goal of conserving energy without unreasonably restricting the freedom of the user in selecting the particular ways she or he chooses to organize the building. Proponents of specification-type controls (such as ASHRAE 90-75) feel that by providing specific restrictions and controls to be followed, there is a greater likelihood that objectives will be quickly achieved. It would appear that both budgets and specification-type controls are likely to be with us for quite some time, including new regulations concerning Building Energy Performance Standards (BEPS).

MONITORING OF COMPLETED INSTALLATIONS

The proof of the pudding, in the case of conservation effectiveness, lies in the fuel bills which are received after the work has been done. Records should be kept, not only of fuel consumed but also of the weather (degree-days), occupancy, and operation patterns, since all affect consumption.

Should actual usage differ from budget, attempts should be made to break down energy consumption by different categories within the structure. For example, GSA estimates energy consumption in the average office building as follows:

Lighting	25%
Heating	35%
Air conditioning	35%
Orther	5%
Total	100%

The Department of Commerce estimates that electric consumption by the average retail store is divided as follows:

Incandescent lights	22%
Fluorescent lights	33%
Air conditioning	32%
Air circulation fans	7%
Other	6%
Total	100%

For more complex buildings, particularly commercial and industrial plants whose operational equipment (rather than the building itself) are heavy energy users, it may be necessary to install a series of devices to measure the flow of fuels, process steam, cooling water, condensate, waste liquids, hot air, process fluids, electricity, and all energy-bearing media throughout the plant.

Where measurements of this complexity are needed, most architects will be well advised to avail themselves of outside consultants for this phase of the work.

Chapter 4

Building Design: General

We made passing reference (Chap. 2) to the ways in which historic architectural design reacted to its environment without reliance on mechanical assistance which, in those early times, was not available. These design considerations included:

- *Siting and orientation:* The relation of the building to the land, the sun, and the wind.
- *Building shape:* The less skin, the less exposure.
- *Nature of the envelope:* Fenestration, insulation, thermal mass, wall shading, color and reflectivity, openings and penetrations.

Let us examine these elements in greater detail.

THE SITE

Site Selection

In actual practice, few architects are offered the luxury of being able to choose among sites in different parts of the country. Selection options, if there are any at all, are usually limited to a few sites within a given community. Most commonly, the client has already selected the site before coming to the architect to discuss the propsed building; and the design consideration becomes one of developing the site and the building as harmoniously as possible to minimize ultimate energy consumption.

Site Development

The two primary energy considerations in the siting of a building are orientation to the sun and orientation to the wind. Landscaping can also improve performance; shade trees can seasonally control direct radiation from the sun; ground surfaces can control reflected radiation; planted ground cover can moderate air temperature; and windbreaks can diminish the force of the wind.

Elements of Site Control

The purpose of site control is to modify adverse climatic forces at a distance before they impact the building. The elements of site control include windbreaks, shade trees, ground surfaces, orientation to the sun and to the wind, and underground structures.

Windbreaks

Windbreaks consist of either a fence or a row of trees or shrubs which reduce air infiltration through windows by diminishing the wind pressure. The most effective location for a windbreak is upwind a distance of 1½ to 2½ times the height of the building. See Fig. 4-1. At this distance, the wind will be deflected up and well over the building, reducing the pushing action on the building's windward side and the pulling action on its leeward side.

A windbreak is more effective if it allows part of the wind to penetrate. A solid windbreak creates a low pressure area on its leeward side with resulting strong eddy currents. These may be as destructive as direct wind in eroding the still air at the surface of the window. Allowing a portion of the wind to pass through the windbreak tends to relieve this leeward suction, as shown in Fig. 4-2.

By protecting the window from the scouring action of the wind, the still layer of boundary air outside the window is preserved, and heat loss is reduced as shown in Fig. 4-3, which illustrates changes in the film factor with different wind speeds.

It should be noted that in many parts of the United States prevailing winter

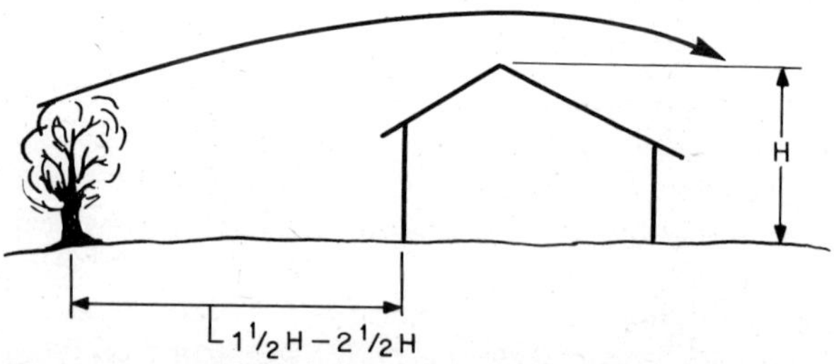

FIG. 4-1 **Windbreak distance.** *(NBS Building Science Series 104, "Window Design Strategies to Conserve Energy," Department of Commerce, 1977.)*

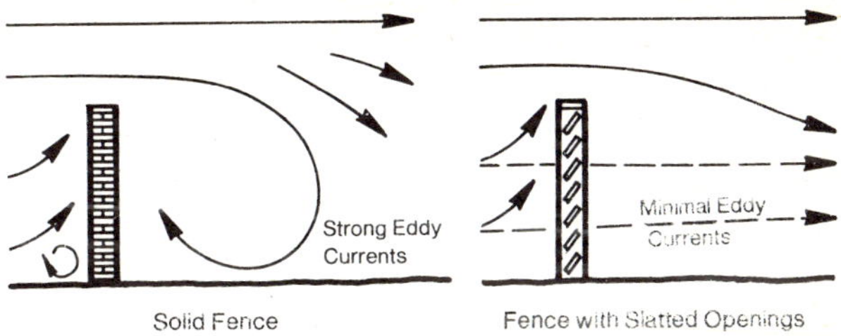

FIG. 4-2 Airflow versus fence design. *(NBS Building Science Series 104, "Window Design Strategies to Conserve Energy," Department of Commerce, 1977.)*

winds come from a different direction than summer winds. Therefore in these locations a winter windbreak need not necessarily interfere with summer breezes.

In an experiment conducted by the Lake State Forest Experimental Station in Nebraska, one house was fully exposed to the wind while an identical house was protected by dense shrubbery. The protected house used 23 percent less fuel to maintain the same environment.

In a South Dakota study, a fully exposed, electrically heated house required 443 kWh to maintain an inside temperature of 70°F from January 17 to February 17. An identical house sheltered by a windbreak required only 270 kWh. The difference in average energy requirements for the whole winter was 34 percent.

Shade Trees

Deciduous trees provide shade in the summer and admit sunlight in the winter. Evergreens provide shade in the summer and reduce window heat loss to the night sky in winter.

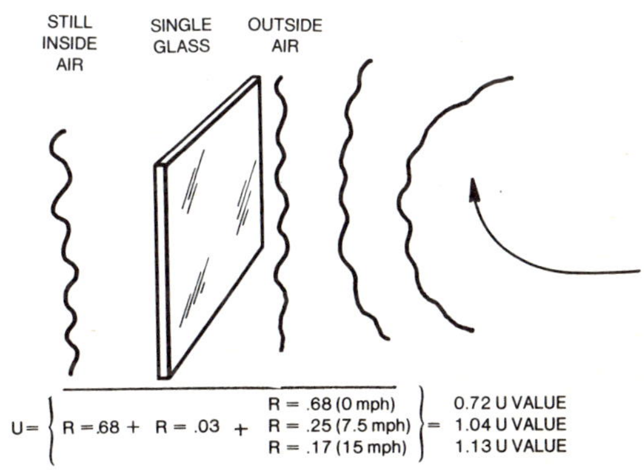

FIG. 4-3 Heat flow through glass versus wind velocity. *(NBS Building Science Series 104, "Window Design Strategies to Conserve Energy," Department of Commerce, 1977.)*

Table 4-1 Light penetration through trees

Type of Forest	Percentage of Light Penetration		
	April	May	Sept.
Evergreen	8	7	4
Deciduous	51	23	5

Source: "Window Design Strategies to Conserve Energy," NBS Building Science Series 104, Department of Commerce, 1977.

A south-facing window shaded by a deciduous tree receives less solar heat than an unshaded north-facing window (the north window receives diffused radiation from clouds). This solar protection increases continuously throughout the summer, as shown in Table 4-1. Trees reduce window heat gain not only by blocking direct sunlight penetration but also by lowering the ground surface temperature, as, for example, when they shade an adjacent parking lot. Placement of the trees should permit winter penetration while blocking summer heat, as shown in Fig. 4-4.

Ground Surfaces (Paved and Planted)

This option involves the use of light-colored ground surfaces to reflect sunlight onto a building, dark-colored surfaces to absorb sunlight and raise outside temperatures, and planted surfaces to both absorb sunlight and lower outside temperatures.

Light reflected from the ground represents 10 to 15 percent of the total daylight transmitted by a first-floor window on the sunlit side of a building. Table 4-2 shows the percentages of incident light reflected by some common ground surfaces.

Ground reflected light transmitted through windows strikes the ceiling. This

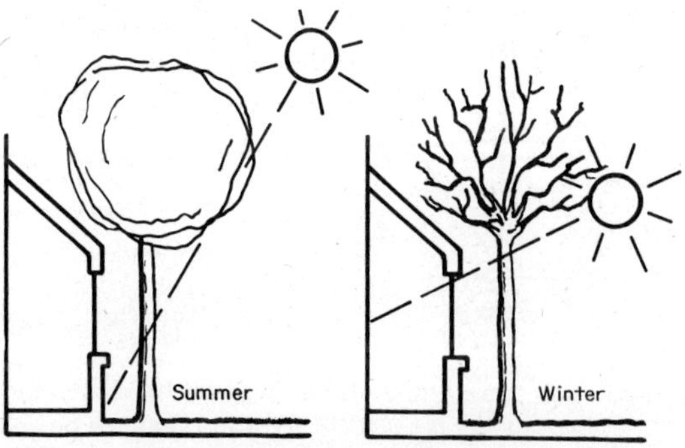

FIG. 4-4 Tree placement. *(NBS Building Science Series 104, "Window Design Strategies to Conserve Energy," Department of Commerce, 1977.)*

Table 4-2 Incident light reflected from the ground

Material	Percentage Reflected
White paint: new	75
old	55
Snow: new	74
old	64
Concrete	55
Marble (white)	45
Granite	40
Brick: buff	48
dark glazed	30
Vegetation (average)	25
Macadam	18

Source: "Window Design Strategies to Conserve Energy," NBS Building Science Series 104, Department of Commerce, 1977.

is beneficial for daylighting in two respects. First, the light is projected deeper into the room than direct sunlight. Second, ceilings are usually light-colored and thus reflect light better than darker floors.

Plant cover absorbs sunlight, yet has a lower temperature than paving because of evaporative cooling which occurs during the transpiration of plants. The net heat gain from the sun is rapidly dissipated by the enormous surface area of the leaves. Very little heat is stored in vegetation because of its minimal mass. Night air temperatures over grass, for example, are therefore cooler than over pavement. The lower day temperatures and lower night temperatures of planted surfaces result in less window heat gain and a reduced air-conditioning burden compared to the situation of having paved surfaces adjacent to the building.

Orientation to the Sun

The general strategy for the placement of windows calls for the largest window area on the side where the sun exposure minimizes combined heating and cooling needs.

Sunlight transmission will be a net benefit on an annual basis if the winter solar heat gain exceeds winter window heat loss and summer solar heat gain. The percentage of the incident solar energy that a window transmits for any given day depends on the angle at which rays of sunlight intercept the window and for how many hours the window receives sunlight.

In general, buildings located in southern latitudes should have window areas concentrated on the northern and southern exposures (ideally with a projecting, horizontal shading device over south-facing windows) to minimize the air-conditioning burden. To obtain the greatest benefit from the sun as a winter heat source, buildings located in northern latitiudes should have window areas concentrated on the south (with minimal window areas to the north).

Table 4-3 Window area (ft²)

	South	North	East	West	Total
Convention	100	100	50	50	300
Revised	180	20	50	50	300

Source: "Window Design Strategies to Conserve Energy," NBS Building Science Series 104, Department of Commerce, 1977.

In a study made for the Boston area, a certain square footage of windows was shifted from the north exposure to the south, as shown in Table 4-3.

The original conventional house was calculated to require 92 million Btu/yr from the heating system. By merely shifting 80 ft² of window from the north to the south, the heating requirement was reduced to 83 million Btu.

Orientation to the Wind

When window placement on opposite sides of an interior space is possible, the building should be oriented slightly askew to the direction of the wind. When window placement on opposite sides of a space is not possible but placement on adjacent sides is possible, the building should face directly into the wind. The reason is illustrated in the plan view of window locations shown in Fig. 4-5.

It will be seen that if the wind encounters an inlet and an outlet in alignment with its outside direction, it will pass through the intervening space in a nar-

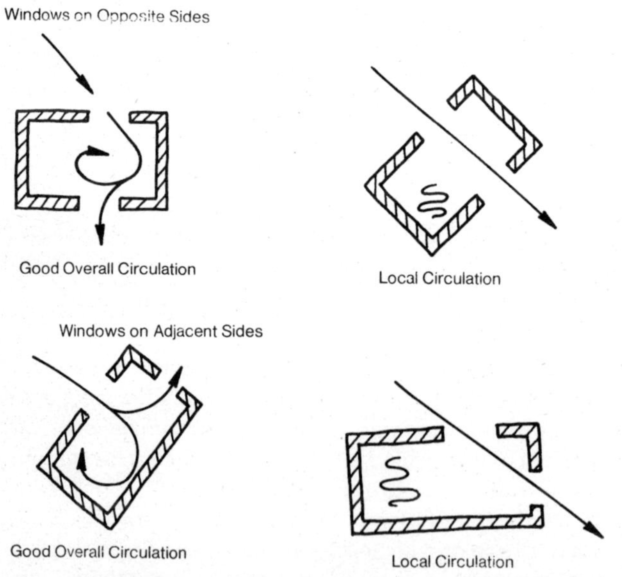

FIG. 4-5 Window location versus air circulation. *(NBS Building Science Series 104, "Window Design Strategies to Conserve Energy," Department of Commerce, 1977.)*

Table 4-4 Effect of window location relative to the wind on the average inside air velocity, given as a percentage of the outside wind velocity

	Wind Perpendicular	Wind at an Angle
Windows on opposite sides	35%	42%
Windows on adjacent sides	45%	37%

Source: "Window Design Strategies to Conserve Energy," NBS Building Science Series 104, Department of Commerce, 1977.

rowly defined, high-velocity stream. Very little ventilation will occur beyond that narrowly defined stream. However, if the wind is forced to change direction in transit between inlet and outlet, a turbulence within the room will develop. A circular current will encompass the sides and corners of the room. The maximum airspeed is reduced compared to windows in direct alignment with the wind, but the average velocity of air movement within the entire space will be greater. Overall ventilation is consequently better.

Where the building interior is subdivided into a series of interconnecting spaces, placement of interior partitions can provide the disruption of the otherwise straight path of airflow between upwind and downwind windows.

Table 4-4 illustrates the effect of window location relative to the wind upon the average inside air velocities, given as a percentage of the outside wind velocity.

Underground Structures

Placed between a building and the outside elements, earth slows the heat transfer from one to the other, reduces the temperature difference between exterior and interior, and at the same time protects the building from cold winds and the direct rays of the sun. While earth does provide some degree of thermal resistance, it is not a particularly good insulator. A layer of insulation, therefore, must be located between the building and the earth which surrounds it, or else the earth will act as a heat sink, always drawing heat away from the interior of the building.

Basic precepts to keep in mind when working with underground structures include the following:

- Gentle south slopes are ideal for underground structures. They are easily built into the hill and have south sunlight and positive drainage.
- Avoid low-lying depressions. Heavy cold air drains into them; frost and dampness are more likely in these areas.
- Make sure that surrounding construction (parking lots, septic systems, etc.) do not drain into the site.
- Identify groundwater levels (including seasonal variations) before making decisions on placement and depth.

- Adequate soil percolation is essential, particularly for sunken courtyards and atriums. If there are problems, consider the installation of overflow drains.
- Any structural system can be used, provided it is designed to applicable loading conditions. General rules are 150 lb/ft^2 for roofs with grass cover and 400 lb/ft^2 for roofs when the earth cover is to support small trees. Add snow, water, and pedestrian loads.
- Wall design is generally the same as conventional basement wall or other below-grade construction.
- Place insulation outside the below-ground building structure; this allows the structure to serve as a heat storage mass. Insulation can be reduced in thickness as the depth below grade increases.
- Butyl sheets provide both waterproofing and a vapor barrier.
- For earth berming against existing exterior walls, a cement plaster finish applied to metal lath and a vapor barrier on the existing wall are advisable. This detail discourages roots, insects, and rodents from getting into the wall.
- Avoid curbs or parapets to retain earth covering. Freezing and thawing action will tend to crack these elements.
- To control interior dampness, dry the air through circulation and/or dehumidification.
- Earth pipes can be used to provide natural cooling. Air taken from the outside during warm weather can be cooled by being passed through long pipes buried in the berm or below grade. The same piping can provide some degree of preheating of fresh outside air during cold weather.
- Examine local codes, especially in relation to fire exits and ventilation.
- Lighting usage should be thought out carefully. It will affect both comfort and energy use more critically in an underground structure.

BUILDING CONFIGURATION

The overall shape of a building affects the amount of energy it will consume. In general, a configuration that resists unwanted heat gains or losses results in an energy-conserving building. To minimize heat transmission for a given enclosed volume, a building should be constructed with minimum exposed surface area. A spherical or round building has less surface and thus less heat gain or loss than any other shape for an equal amount of total floor area. A square building has less surface than a rectangular one of equal area per floor, and so undergoes less thermal transmission loss or heat gain. However, the number of stories modifies this relationship for the building as a whole.

The exception to this generalization about a building's configuration is the situation where the primary thermal load is the result not of the environment, but rather of internally generated conditions. Where an office building, for example, generates large amounts of heat from its lighting and computers, the problem may be one of dissipating rather than conserving thermal energy. In this case it may (depending on the climate) be desirable to have as large an amount of building skin as possible to facilitate heat dissipation.

In its "Energy Conservation Design Guidelines for New Office Buildings" GSA adds the following observations with respect to building configuration:

1 *Tall Buildings:* A tall building has a proportionately smaller roof and is less affected by solar gains on that surface. On the other hand, tall buildings generally are subjected to greater wind velocities, which increase infiltration and heat losses. Tall buildings are less likely to be shaded or protected from winds by surrounding buildings and trees. They require more mechanical support systems, including elevators and longer exhaust duct systems. The stack induction action in tall buildings increases infiltration, thus requiring special measures to reduce its influence on heat gain and heat loss.

2 *Floor-to-Ceiling Height:* Greater ceiling heights improve environmental conditions in the summertime by permitting warm air to rise. However, greater ceiling heights increase the perimeter areas, thus increasing heat transmissions through the walls.

Reduced ceiling heights reduce the exposed exterior wall surface area and the enclosed volume. A reduced ceiling height can also increase illumination effectiveness.

Floor-to-ceiling height is determined by physiological comfort, height of light fixtures for proper light distribution, and height of windows necessary for good natural lighting.

In general, increases in ceiling height need increase only the exposed wall surface (not window surface). The effect of greater heights on energy consumption may be rather small, depending on the thermal characteristics of the wall.

3 *Ceiling Plenum Height:* Deep ceiling plenums allow the use of larger duct sizes with low pressure drop and reduced heating, ventilation, and air-conditioning (HVAC) air handler and fan requirements. Ducts can be larger, allowing greater volumes of air to be moved with smaller pressure drops, permitting reduced fan horsepower.

4 *Roofs:* Very low buildings may have greater roof area in proportion to wall area, and the heating and cooling loads which they generate may, in turn, influence the selection of the mechanical equipment. In tall buildings, the roof is a lesser influence on the total heat loss and gain, and will rarely influence the selection of the total heating and cooling systems.

5 *Exposed Floors:* Buildings that are elevated on columns or with small first-floor areas and large overhanging upper floors increase heat loss and heat gain because of the extra exposed floor surfaces. While this may be of slight advantage all year in southern regions, or anywhere in the summertime, it presents a serious increased heat loss in colder climates. Locating parking garages on intermediate levels similarly increases energy consumption from additional exposed surfaces.

6 *Building Forms:* A dome roof can permit warm air to rise and collect at the top, leaving the floor area cooler. Pyramids, zigzag exterior walls, rhomboid-shaped buildings, and other forms can all be used to control the influence of climate on consumption.

7 *Zigzag Walls:* Zigzag configuration of east and west walls provides self-shading to reduce summer solar loads, provides natural windbreaks, and can permit low rays to penetrate the building in the winter to supplement the heating system, if the windows in the zigzag are facing south. By facing the windows north in the zigzag in a southern location, heat gain is reduced year-round; but, in both summer and winter, natural lighting and views can be available at both east and west facades without the penalty of increasing summer heat gains. However, the energy requirements resulting from the additional wall surface for the zigzag form must be weighed against the other energy benefits.

PASSIVE SOLAR DESIGN

As a result of the return to traditional principles of environmental design of buildings, we are currently witnessing escalating interest in various aspects of passive solar design.

Initially, our national effort to spur solar design was aimed at the so-called active systems, which used various pieces of equipment and control devices to achieve their end. This included solar collectors, storage systems, and distribution systems, all of which were added to the building.

Recently a shift to the passive approach aimed at the same goals of heating or cooling by relying on little beyond the building itself. Some solar faddists were quite strict in their refusal to use anything other than the basic building, and have been willing to accept rather broad "comfort" extremes in order to avoid the use of supplementary equipment.

To the average architectural practitioner, however, a broader and perhaps more appropriate definition would permit the use of intermittent inputs of nonrenewable energy, generally in the form of electricity, to operate control devices, to move insulation or water, etc.

In general, modern passive design takes one of three approaches:

Direct Gain: South wall or clerestory windows
Shading overhangs for summer
Internal mass

Indirect Gain: Thermal storage wall
Thermal storage roof
Solar greenhouse
Natural convective loop

Isolated Gain: Indirect gain situation in which there is a major separation (by either distance or insulation) between thermal storage and conditioned space

This is not a text on solar energy, but many of the elements in the following section conform to principles of passive solar. Indeed, any serious student of energy conservation in buildings should keep abreast of continuing developments in this area. Many years from now, when the final accounting is in for the whole solar energy effort, it may well turn out that results from the passive effort will outweigh those of the active procedures which currently attract most of our attention.

BUILDING ENVELOPE

A building's exterior envelope transmits and stores the thermal energy passing between the structure and its environment. Two aspects of envelope composition are receiving increasing emphasis in the effort to create energy-conscious design. These are, first, the change from static envelope design to the concept of a modulated construction and, secondly, greater consideration of the influence of thermal mass on overall energy consumption.

Modulated Construction

Traditionally, walls and roofs have been thought of as permanent, fixed elements of construction. While energy-conscious design increasingly calls for a different treatment of each facade based on its orientation, each facade is built so that its face to the weather remains unchanged both during each day and during different seasons.

The concept of modulated construction is based on the supposition that just as different orientations require different treatments, so might different times of day, or different times of year. Looked at from this point of view, the building envelope should respond to the change from day to night in the same way as a flower might open itself up to the sun and then close itself for protection against the cool of the night.

Recent developments in passive solar design have suggested different ways in which this modulating, flexible concept might be incorporated into the envelope. Flexible wall possibilities include:

- Walls which slide, horizontally or vertically, fully opening up the facade when desired.
- Walls which slide over glass areas in order to reduce heat loss in winter and heat gain in summer.
- Walls which revolve or rotate in order to expose an appropriate material to the sun or to the inside.

Traditional window shutters, sliding insulated panels, and storm windows and doors are among traditional ways of making these adjustments.

The Beadwall® concept developed by Zomeworks, Inc., of Albuquerque, New Mexico, changes the insulating characteristics of a glass wall by constructing the window of two glass sheets separated by a 3-in airspace. As shown in Fig. 4-6, this space is filled with styrofoam beads blown into the cavity by a small pump and blower system when increased insulation is needed at night. Similarly, the beads are evacuated by the system when the structure should be open to receive solar energy during the day.

Operable skylights offer opportunities to modulate the roof in order to improve its responsiveness to energy needs. In another Zomeworks concept (Fig. 4-7), their Skylid® allows penetration of light and heat when desired, but it automatically closes when conditions suggest; canisters of Freon gas are used to provide the controls.

Still in the research and development stages are materials which will change their structure (and hence their thermal response) as conditions change in the environment inside and outside the building. This is somewhat like the new type

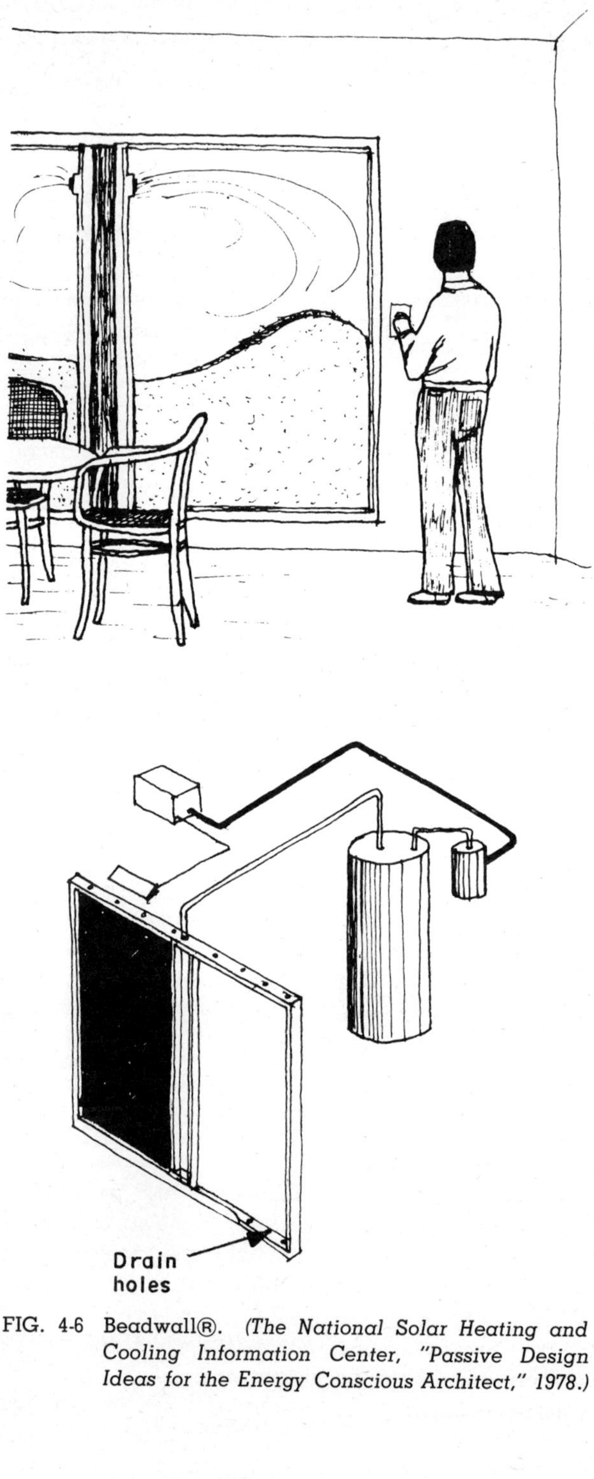

FIG. 4-6 Beadwall®. *(The National Solar Heating and Cooling Information Center, "Passive Design Ideas for the Energy Conscious Architect," 1978.)*

FIG. 4-7 Skylid® by Zomeworks, Inc. *(The National Solar Heating and Cooling Information Center, "Passive Design Ideas for the Energy Conscious Architect," 1978.)*

of sunglasses which darken automatically as the light becomes brighter and then become more transparent when the light fades. In similar fashion, the insulating ability of these new envelope materials will vary in response to changes in temperature. When they are fully developed, the hope is that envelope modulation can take place automatically within the structure of the envelope material and thus make unnecessary the physical changes such as shutters, beadwalls, etc., which are our only tools at present.

Thermal Mass

Mass, the property of density, can have a great effect on the energy consumption of buildings, and is sometimes of greater consequence than insulation. Mass has the ability to store heat and thus to delay the effects of the outdoor temperature cycle. Old buildings of heavy masonry construction are good examples of the moderating effect of heavy construction, whereas lightweight construction of wood or metal does little to delay heat flow, thus making it necessary to restrict the flow of energy with insulation. By adding enough insulation, the energy flow can be reduced to the point where very little heat can get in or out.

The concept can be illustrated by imagining a structure that is modeled from sponges, wrapped in paper, and submerged in a tub of water. The sponges represent its mass, and the paper its insulation. The water represents the heat which surrounds the structure. The paper restricts the amount of water which can enter the building but does permit some of it to seep through. The sponges have the ability to absorb a quantity of water before they become saturated and allow the water to penetrate the structure.

Every building material has the ability to both resist and delay the flow of energy. Some materials are better in one characteristic than in another. Brick and concrete can absorb large quantities of heat, but do little to resist flow. Conversely, fiber glass insulation will not conduct heat well because of its high resistance, but cannot absorb or delay the passage of energy.

Resistance is defined as the reciprocal of the amount of Btu's which will be transferred through one inch thickness of material one square foot in area in one hour when the driving force is a one degree Fahrenheit temperature differential.

The *energy capacity* of a material is the number of British thermal units it takes to raise a one inch thick by one foot square of a given material one degree Fahrenheit. Listings for the resistance and energy capacities of common building materials may be found in the Department of Energy publication "Minimum Energy Dwelling (MED) Workbook."

In certain situations both insulation and mass excel in the conservation of energy. But if used improperly, each can actually increase consumption.

To repeat, the value of insulation is its ability to restrict the flow of energy, and the value of mass to delay the flow.

Insulation is very important in climates with extremes in yearly temperatures. A good example is the North Central United States. In the winter it can become extremely cold, and the summer can be extremely hot. A building in such a climate needs large amounts of insulation to restrict the flow of heat in both winter and summer. But in a moderate climate such as San Francisco, excessive insulation in an office building may actually be detrimental because it retains internally generated heat (from lights, people, computers) rather than letting it leak out. Under these circumstances, the insulation actually results in the need for more cooling energy. It is always important to determine in which direction heat flow will occur, for what periods of time that flow will occur, and whether the flow is detrimental or beneficial before the decision is made concerning the amount of insulation which should be used.

The best use of mass is in buildings which have large daily swings in energy consumption. The best possible situations are buildings which require equal amounts of heating and cooling over a 24-hour cycle. By delaying the heat buildup during the day, the heat can be used to warm the building at night.

Even buildings which require only heating or only cooling may find mass worthwhile. By damping the extremes of temperature which impact on the interior, mass permits the use of smaller mechanical equipment (thus lowering initial cost) and makes it possible for the equipment to run more continuously and thus more efficiently (and by so doing to reduce operating cost).

Mass, like insulation, can be a disadvantage if used improperly. A building with infrequent and changing use patterns such as a church or a theater is a good example. A church may be used for only 4 hours on the weekend and not require heating to the comfort zone for the remaining 164 hours in the week. Two hours before the church service, the lights are turned on and the thermostat is turned up. If the building is of massive construction, it may remain cool throughout the duration of the church service only to become comfortable after everyone has gone.

Mass is slow to respond to controls, which must be able to anticipate conditions hours in advance. Factors to be considered include the general climate conditions, diurnal swings, and occupancy modes.

Placement of insulation is also a factor in the amount of time needed for thermal energy to work its way through a given mass. Consider, for example, 2 in of rigid insulation and 8 in of concrete block. Separately, the insulation has a conductivity "U" value of 0.18 and a time lag of 0.75 h. The concrete block has a "U" value of 0.5 and a time lag of 4 h. If the insulation is placed on the interior of the block, the sandwich has a time lag of 8 h. If the insulation is placed in the middle of the block, the time lag becomes 10 h. If the insulation is placed on the exterior, the time lag becomes 13 h.

The placement of insulation has such an effect because of the movement of heat into and out of the wall construction. In a homogeneous wall, the heat flows in and out are equal. But within a wall with exterior insulation, for example, the heat flow becomes more selective. The insulation restricts the flow into the massive heat-absorbing material, and therefore more time elapses before the storage capacity of the massive material is filled, increasing the time lag. The increase of time lag resulting from changing the insulation placement in the previous examples is 5 hours, while at the same time the U value of the wall remains unchanged.

The ultimate in mass is a dwelling built in a cave. The time lag becomes infinitely large, and the indoor temperature is little affected by the diurnal cycling of the outside air temperature. The dwelling remains at a constant ground temperature during the day. In practical terms, the ultimate uses of mass are difficult to achieve at low cost. Mass, like insulation, has an economic optimum where the benefit of improved performance is not worth the additional first cost.

A future possibility of increasing thermal mass lies in the use of heat-of-fusion chemicals as thermal storage devices. The heat of fusion is the amount of energy required to change the state of material from a solid to a liquid. The amount of this fusion energy is considerably greater than that required by a simple rise in temperature. For example, the amount of energy required to heat 1 gal of water 1°F is approximately 8 Btu; but if the water is frozen at 32°F, it takes 143 Btu to raise it 1°F or 18 times the amount of energy. Thus the use of the heat-of-fusion concept allows the use of much smaller amounts of materials.

A recent development uses a heat-of-fusion material which has a melting point of 74°F. The concept is a solar tile. The heat-of-fusion material consists of a 2 ft × 8 ft panel, ½ in thick, subdivided into ¼-in segments by a fiber glass membrane. Two panels are then wrapped in a polyester concrete to produce a 4 ft × 8 ft building panel. The total building material is less than 2 in thick but has the heat capacity potential of a 10-in masonry wall. These developments are still in early stages, but may someday impact the marketplace and building design.

Chapter 5

Building Design: Components

The last chapter dealt with overall building design. In this chapter, additional emphasis is placed on certain building components for which greater detail is desirable. These include window treatment; insulation; solid waste; heating, ventilating, air conditioning; domestic hot water; electric power; lighting; vertical transportation; and operation and maintenance.

It will be noted that many of these elements are mechanical in nature and, as such, will be handled by the architect's mechanical consultants. Accordingly, the discussion of these items is intended to provide a general background against which the architect can ask appropriate questions to verify that consultants are, in fact, doing their job.

WINDOW TREATMENT

Well-designed windows can actually provide a net energy gain; when poorly designed, they can be an enormous energy burden. Basically, windows have five energy functions:

- Providing winter solar heat
- Providing year-round daylighting
- Rejecting summer solar heat
- Providing insulation and airtightness during periods of heating or cooling
- Providing natural ventilation during temperate weather

Figure 5-1 provides an example of how sunlight passes through ordinary glass. The numbers in the figure are subject to great variation, particularly the percentage of incident light that is reflected from the outside surface of the glass. As the incident light strikes the window more obliquely, penetration through the glass drops off sharply. This feature is helpful because it reduces solar gain when the sun is at higher elevations. Once light of any wavelength has passed through the glass and has been absorbed in the interior of the building, the energy is trapped, because glass is opaque to the infrared radiation emitted by bodies at room temperature (the so-called greenhouse effect).

To enhance the thermal effectiveness of windows, a variety of treatments has been developed over the years. These fall into two categories: manipulating light before it hits the window and changes in the elements comprising the window. Light manipulation before it hits the window is accomplished primarily through a variety of shading devices consisting of either overhangs above windows or louvers.

In the case of overhangs, the objective is to design the overhang in such a way that the window beneath it is adequately shaded as a result of the high elevation of the sun in summer, while sunlight is allowed to enter in the winter. The azimuthal angle between the shaded surface and the sun is an important factor in the effectiveness of the shading device. For a given elevation of the sun, shading devices provide least coverage when the sun is perpendicular to the shaded wall. As the sun moves to the side, the shadow of the shading device lengthens, increasing its coverage. For this reason, horizontal shading is particularly effective on walls facing southward, because the lengthening shadow of the shading device compensates for the lower elevation of the sun when it is in the East or West.

There are a variety of louver materials. One variety consists of aluminum

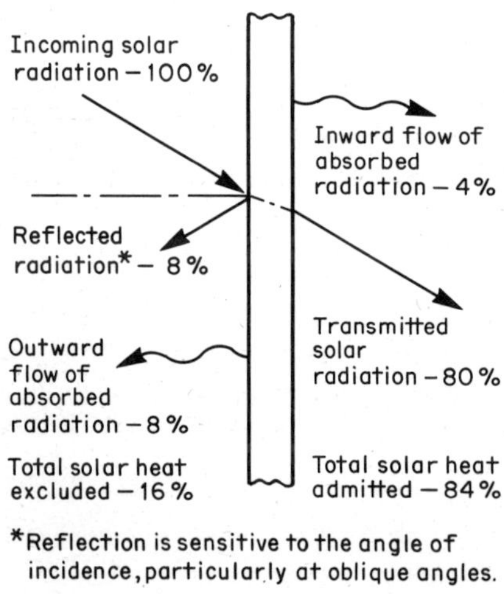

FIG. 5-1 Typical light transmission of clear glass. *(ASHRAE Journal, November 1978.)*

sheet metal embossed to form small louvers, typically several millimeters apart. Another variety consists of small bronze strips held together by vertically woven wires and mounted in frames. All types are mounted on the building exterior. Because the louvers admit substantially more sunlight when the sun is at lower elevations, they admit more solar energy in winter than in summer and thus are more suitable in northerly locations than nondirectional meshes.

A number of exterior devices are available to supplement overhangs and louvers. These include exterior roll blinds which can be raised or lowered, exterior shutters which can be opened or closed, and awnings made of a variety of materials.

In considering changes in window construction, every element is susceptible to change. In the case of glazing, options include multiple glazing, reflective glass, heat-absorbing glass, and reflective and absorptive films.

Multiple glazing, as either multiple-pane windows or storm windows, is among the most common solutions to the very poor insulating characteristics of glass.

Reflective glass solves another part of the sun control problem—the rejection of excessive solar radiation. While reflective glass can be of considerable advantage during the cooling season, it also rejects winter heat from the sun which might be desirable. Accordingly it is absolutely necessary to perform an analysis of the effect of reflective glass over an entire yearly cycle of weather.

Heat-absorbing glass is capable of absorbing selected wavelengths of light. Because heat is absorbed into the glass, rather than being reflected, the glass itself will become warm. In more northerly climates, this may be a desirable way of limiting illumination levels while retaining some of the heating qualities of sunlight. Manufacturers of window glass offer a wide range of combinations of reflective and absorptive properties, and it is important that the proper combination be used.

Reflective and absorptive films offer the same energy conservation properties as reflective and absorptive glass. Films, which are applied internally, can be purchased in many combinations of reflectivity and absorptivity. Their principal disadvantage is the fragility of their plastic surface and uncertainty as to service life.

Changes in the construction of the window frame in response to energy considerations have resulted in frames which are built of materials that are good insulators or, in the case of metal construction, which embody a thermal break aimed at stopping conduction losses between interior and exterior. Other areas of exploration include a frame design in which ventilation takes place through the frame itself through a controlled opening, while the glazing remains fixed; the result is a marked reduction in crack loss infiltration.

Finally, though least effective in terms of control, various interior accessories are used. These include venetian blinds, draperies, film shades, opaque roll shades, and insulating shutters.

The preceding discussion was concerned primarily with the thermal impact of windows. The proper use of windows for daylighting, in lieu of artificial lighting, is of growing importance in the conservation effort.

INSULATION

Architectural aspects of different types of insulation include.

- Physical structure
- Moisture transmission
- Fire resistance
- Installation
- Available sizes

Aluminum Foil

- Aluminum foil is single or multiple parallel layers of foil and kraft paper combinations with an airspace. Being a metal, aluminum foil is a good reflector but a poor absorber of radiation. It reflects 95 percent of the heat and absorbs only 5 percent.
- Aluminum foil is almost impervious with a vapor barrier rating of 0 to 0.02 perm.
- Aluminum is fire-resistant, but kraft paper laminate is combustible.
- It is easily stapled in place.
- It comes 25 in wide for rafters, joists or studs on 24-in centers; 33 in wide for 16-in centers; and 36 in wide for 16-in centers where insulation is to be recessed.

Urethane

- Urethane is cellular plastic foam formed by the reaction of isocyanates and polyols. As it foams and expands, it entraps fluorocarbon vapor which has extremely low heat conductance.
- Vapor transmission is 2 to 3 perms. It is not a vapor barrier, and a separate one should be specified.
- Urethane is combustible, and it must be protected from fire by finish surface materials. Do not use at temperatures above 200°F.
- Foam application requires special machines and skills. Urethane board is light and easily worked.
- Foam expands to fit any shape or size space. Tapered roofing sections are usually 2 ft × 4 ft, or 3 ft × 4 ft boards tapered 1/16, 1/8, or 1/4 in.

Expanded Polystyrene

- Expanded polystyrene is a polymer impregnated with a foaming agent which, when exposed to heat, creates a uniform closed cell structure highly resistant to heat flow and moisture penetration.

- It resists moisture penetration.
- Expanded polystyrene is combustible and must be protected by finish materials.
- This material is easily cut and lightweight.
- It comes in boards in lengths of 4, 6, and 8 ft; widths of 12, 16, 24, and 48 in; thicknesses of ½ to 19 in; tapered roof insulation.

Extruded Polystyrene

- In composition extruded polystyrene is similar to expanded polystyrene except the extrusion process compresses the cells, leaving no voids for better moisture protection.
- It is inherently resistant to all forms of moisture.
- It contains a flame retardant but is still considered combustible and must be protected.
- Install with mastic or using self-tapping screws into metal studs.
- It comes in 8-ft boards, 24 in wide, ¾, 1, and 2 in thick. Tongue and groove sheathing 8 ft long; 16 or 24 in wide; ¾, 1, and 2 in thick. Roofing boards are 24 in × 48 in; 1, 1½, 2, and 3 in thick.

Cellular Glass

- Cellular glass is a lightweight, rigid insulation composed of millions of completely sealed glass cells.
- It is completely impervious to moisture.
- It is totally inorganic and noncombustible.
- Adhesive is used to hold glass board to wall surfaces.
- It comes in 24 in × 48 in boards in 1½ to 4-in thicknesses; 12 in × 18 in and 18 in × 24 in flat blocks in thicknesses from 1½ to 4 in; tapered roof blocks 18 in × 24 in with tapers of 1/16, ⅛, or ¼ in.

Fiber Glass

- Fiber glass is made of fine woollike compressible fibers spun from glass. It can be in loose form, in batts, or faced on one or both sides with paper or foil.
- Moisture transmission of foil-faced fiber glass is 0.5 perm; kraft-faced 1.0 perm. Unfaced should be protected with vapor barrier.
- Fiber glass is noncombustible, but its backing is not.
- The loose type is blown in. Most typical are batts which are stapled to framing or held in by friction fit.

• There are a great variety of shapes and sizes including batts in 16 or 24-in widths 4 ft long, and 3½, 6, or 9-in rolls of 32-ft lengths. Use sill sealer of 1 in × 6 in × 50 in. It comes loose in 4-ft³ bags.

Urea Formaldehyde

• Urea formaldehyde is a low-density foam that can flow into odd-shaped spaces. There is some shrinkage when it is cured.

• It should be protected by a vapor barrier.

• It is noncombustible, but its toxicity should be investigated.

• Foam is cold-setting in 10 to 60 s. It can travel before setting and requires special equipment by contractor.

• Foam spreads to fit any shape or size space.

Cellulose

• Cellulose is wood fiber insulation made from recycled newsprint in loose form, sprayed, or as boards.

• It should be protected with a vapor barrier.

• It can be treated to be rated noncombustible but should not be used at temperatures over 150°F.

• It can be hand-poured or machine-blown; it can also be sprayed.

• Cellulose comes in 4-ft³ or 15- and 20-lb bags.

Mineral Wool

• Mineral wool is made of woollike fibers spun from molten slag and coke, similar to fiber glass.

• It transmits moisture easily.

• It will not burn or melt at temperatures below 1800°F.

• Mineral wool can be hand-poured or machine-blown or sprayed. Batts are stapled or friction-fit.

• It comes in 26-lb bags; batts with or without foil backing in 3⅝, 6, 6½-in thicknesses, 16 and 24-in widths, 48-in lengths.

Vermiculite

• Vermiculite insulation is made of lightweight granules of free-flowing vermiculite, a clay mineral consisting of mica.

• Particles resist moisture infiltration, but vapor barriers should be used in high-humidity situations.

- It is noncombustible.
- It is easily poured.
- It comes in 4-ft^3 bags.

Perlite

- Perlite is composed of lightweight granules of inert volcanic glass expanded by a heat process and treated with a nonflammable silicone.
- Perlite particles are waterproofed but cannot retard heavy infiltration.
- It is noncombustible.
- It is easily poured.
- It comes in 4-ft^3 bags or 1- and 2-in rigid boards.

SOLID WASTE

The amount of solid waste generated within a building varies, depending on the building's use. Common estimates are that an office building generates 1 or 2 lb of solid waste per person per day; a residence, 5 or 6 lb per person per day; a hospital, as much as 20 lb. Half of solid waste, on the average, is paper and cardboard. This, along with plastics, wood, ferrous metal, aluminum, glass, tar, naphtha, textiles, oil, ash, and foodstuffs, is theoretically available as fuel in some kind of a recycling operation.

Generally, however, where off-site facilities exist for solid-waste disposal, it is cheaper to use them rather than undertake the construction of one's own recycling operation. Two exceptions to this generalization are an industrial plant which produces large and continuous quantities of a waste product that can be burned without expensive preliminary separation processes, and a community recycling plant receiving waste from a number of neighborhood buildings. For example, in Nashville, Tennessee, a $16.5 million solid-waste fuel plant is lowering the costs of heating and cooling a number of high-rise buildings. Twenty-seven buildings are hooked into the system, which will pipe high-pressure steam for heating and chilled water for cooling at 15 to 20 percent savings for building owners, compared to conventional systems. Owners can cut capital costs because they can eliminate heating and cooling plants for their individual buildings. The system uses natural gas and oil as standby fuels and is estimated to reduce air pollution emissions from the participating buildings by 75 percent.

HEATING, VENTILATION, AIR CONDITIONING

This phase of conservation will almost always be performed by the architect's mechanical consultants. Nontheless, the architect can play an important role in the basic programming of the structure. This programming has two main areas:

1 Planning the building so that as large an area as possible can be shut down during off hours, such as nights and weekends, with a consequent reduction in energy use.

2 In conjuction with the owner and the mechanical engineer, setting up criteria for the design of systems within code limits so as to permit the most efficient selection and operation of equipment. For example, designers have commonly used criteria calling for 1 or 2.5 percent weather conditions (design based on projected weather conditions exceeded only 2.5 percent of the year).

Today this practice is being questioned. Since 2.5 percent conditions are peaks rarely reached during occupied hours, it is often wasteful to design a building as if it were always going to operate at that level. The resultant oversizing of equipment brings not only higher initial costs, but also the intermittent operation that means higher energy use. Accordingly many designers are considering 5 or 10 percent conditions as a more realistic objective, provided the client is willing to undergo occasional discomfort.

Similarly, more realism in assumed hours of occupancy can result in large savings. For example, in summer an office building need not necessarily be completely cooled and ready to go when the very first workers start to arrive. Instead the cooling might be scheduled to be online at 10 a.m. rather than 8:30 a.m. At the end of the day, shutdowns that begin at 4 p.m. rather than 5 p.m. can bring corresponding savings. There is obviously no way to codify a single solution for all situations but the architect should make a strong effort to see that discussion involving criteria for mechanical design is given enough time in order to avoid ending up with a system that will always consume more energy than necessary.

The use of the modular approach to equipment selection will continue to increase. For example, the use of several smaller boilers instead of a single large one allows some of the smaller ones to run at full capacity—and thus efficiency—for most of the time. In contrast, a single, large unit must cut in and out in response to its ability to satisfy most demands quickly and thus usually operates with lower seasonal efficiency. In selecting modular units, discussion involving space allocation for the smaller units should be started early. As designers acquire more experience with the performance of modular systems, we will probably see some reduction in the number of small units initially installed, but with space being left for the addition of others, should actual operation indicate the need. Experience with most modular installations to date indicates that times when all units are needed are few and far between—if ever.

Particulars relating to the design or retrofit of HVAC equipment fall to the architect's consultants and are not within the scope of this text. Conservation expertise is increasing in all areas of HVAC design and is in a constant state of flux. As always, the selection of a competent consultant is of the greatest importance.

DOMESTIC HOT WATER

Energy conservation through efficient domestic hot-water systems is most significant in hospitals (18 to 30 gal per bed per day), homes (20 to 25 percent of total fuel used), and other facilities which frequently use large amounts of water. At the other end of the scale are office and most commercial buildings where hot-water consumption is minimal.

The principles for energy conservation in hot-water systems are similar to those for air handling. Water requires pumping; so the less water used, the less energy is required. The process of heating water consumes energy, so the lower the permissible temperature levels, the less energy used. Among the techniques used to cut energy consumption are:

- Quantity and temperature control: Faucets should be equipped with spray heads or flow limiters. Heat faucet water to 105°F maximum. Consider using self-closing valves. Where water hotter than 105°F is required, as for dishwashers and certain equipment in commercial kitchens, use booster heaters located close to the point of use. The energy for these boosters should be either steam or electricity, depending on analysis.
- System Design: If fossil-fuel boilers are the heat source for hot water, provide a separate boiler which matches the hot-water load rather than using an oversized heating boiler all summer. The hot-water circulating system should be designed so that the circulating pump can be turned off during unoccupied hours.
- Heating with waste heat: Domestic hot-water generation at low temperatures is a good application for waste heat energy including waste heat from incinerators, rejected heat of compression from refrigeration units, hot condensate return from steam systems, rejected heat from diesel or gas engines, waste heat from transformers, and heat in toilet exhaust air.
- Solar heat source: In time, with sufficient tax and other financial incentives, solar-heated hot water will play a larger role.

ELECTRIC POWER

Rate Structures

Many utilities have ten or more different rate structures, each often subject to one or more riders. The utility is under no obligation to point out the cheapest schedule, and an analysis should be run to verify that the most economical is being used. The analysis includes:

- Rate schedule and riders used
- Maximum demand and period of occurrence

- Power Factor
- Monthly and annual energy consumption
- Average cost per kilowatthour
- Service voltage level and secondary-use level
- Transformer and equipment ownership

Utility Bills

All utility bills should be reviewed for the following factors.

Energy

Most electric utilities use a sliding block approach for energy charges, that is, so much per kilowatthour for the first 1000kWh, so much per kilowatthour for the second thousand, and so on. In most cases, the more kilowatthours consumed, the lower the rate.

Fuel Adjustment

Many electric utilities have imposed an additional energy charge, called a fuel adjustment, which reflects the increased cost of fuel. It can be a substantial charge. [Gas utilities have something similar called a purchased gas adjustment (PGA) charge which reflects adjustments to gas bills based on fluctuations in the wholesale cost of gas.]

Customer-Related Costs

Some utilities add what is called customer-related costs. These comprise a special charge which reflects part of the distribution investment, part of the operating and maintenance costs, costs for accounting and collection, and so on. If it is not billed separately, the customer-related cost may be included with the energy cost.

Demand

The demand charge is designed to make the customer pay his or her fair share of the utility's fixed investment in the production, transmission, and distribution equipment required to satisfy maximum-load needs. The charge is based on the rate at which electricity is consumed. The more energy used at any given time, the larger the utility's investment in generation, transmission, and distribution systems has to be. For example, consider two users, A and B, both consuming an equal number of kilowatthours each day. User A consumes electric energy 24 hours a day, and user B consumes it 8 hours a day. User B requires the utility to have capacity equal to the ratio of 24/8 times the capacity required to supply user A. This is 3 times the capital investment that was required to serve user A; so user B is billed for this extra investment.

The consumer's actual demand is computed as the average amount of energy consumed in a predetermined demand measurement interval, usually 15 minutes (other intervals are also used by the utilities, particularly 30 and 60 minutes). Regardless of the interval, the highest demand recorded during the month becomes the actual demand billed for that month.

Many utilities also employ a special clause which states that no matter what the actual demand may be in any given month, the demand billed may be no less than a certain percentage of a demand recorded typically. This is known as a "ratchet" charge, and it is intended to reflect the need for the utility to be ready to meet a heavy demand even though it is not actually called upon to do so in that month.

Low-Power-Factor Penalty

Another charge sometimes applied is a penalty for a low power factor. Electric utilities must supply both real and reactive power for their customers. Reactive power does not register on a kilowatthour meter, but producing it still requires the utility to put additional investment into generating, transmission, and distribution facilities.

Electric Power Conservation falls into two categories:

- *Load reduction*: This means designing or redoing a system so that it uses as little power as possible. This involves careful analysis of both electric distribution systems and the electric-using equipment selected for the building.
- *Load management*: This term refers to demand control systems (usually computerized) which shed predetermined loads to keep the demand level within a predetermined maximum and power factor improvement (primarily through installation of capacitors at equipment locations) aimed at improving the power factor.

Available to the architect for this work are not only the electrical engineer, but also specialized utility consultants familiar with local rate schedules and billing procedures.

LIGHTING

One of the most visible sources of energy waste in a building is the traditionally designed lighting system. Excess lighting results in energy waste both for lighting itself and in warm weather for the cooling needed to offset the heat created by the lights.

In low-rise buildings without elevators, lighting can consume 60 percent of the supplied electric power. In hot weather, lighting systems account for one-fourth to two-thirds of the cooling load in a typical office building.

Some aspects of the relation between daylighting and artificial lighting are of interest. Daylighting does not reduce the amount of artificial lighting that must be installed in a building, since it must be designed for the worst, i.e., night condition. It does, however, reduce the total consumption of lighting energy for certain areas which receive sufficient natural light to allow the lights to be turned off for part of the occupied hours. Daylighting requires proportionately more glazing in the building envelope, thus increasing heat transmission and possibly infiltration. Whether the lighting savings as a result of increased glazing is offset by the increase in energy required for thermal purposes as a result of this glazing is a consideration best left to the computers of the mechanical engineers. In any event, areas near windows should have a system of switches that allow the artificial lighting to be turned on and off, or dimmed, according to the amount of daylighting available at a given time.

With respect to artificial lighting, there are basically three approaches to saving energy:

- Reduce lighting levels where practical.
- Turn off lights where possible.
- Upgrade the efficiency of lighting fixtures.

Illumination Levels

One of the most important contributions an architect can make is to design space that is light-efficient. In general, room surfaces should have light colors with high reflectance and matte finishes. The wall colors should have a light reflective value of 60 percent or greater. Ceilings should be white. Desk tops, filing cabinets, and other furnishings within the field of vision should be light-colored with matte finishes. Open planning permits the greatest light flow over a given area.

In most buildings there are three general categories of illumination required: specific task lighting, general lighting around tasks, and general lighting for circulation and support areas. The traditional candlepower recommendations for these functions are being reevaluated and reduced in response to the energy crisis, and the designer should keep current with these changes. Typical of these standards existing at the time this text was prepared (but perhaps outdated by the time it appears) are the FEA recommendations in Table 5-1.

Lights and Switches

A lighting use program should be established to schedule lighting use with building occupancy patterns. Switches greatly affect energy consumption in lighting systems. Contrary to general belief, it is rarely economical in terms of

Table 5-1 FEA-recommended maximum lighting levels

Task or Area	Footcandle Levels	How Measured
Hallways or corridors	10 ± 5	Measured average, minimum 1fc
Work and circulation areas surrounding work stations	30 ± 5	Measured average
Normal office work, such as reading and writing (on task only), store shelves, and general display areas	50 ± 10	Measured at work station
Prolonged office work that is somewhat difficult visually (on task only)	75 ± 15	Measured at work station
Prolonged office work that is visually difficult and critical in nature (on task only)	100 ± 20	Measured at work station
Industrial tasks	ANSI-A11.1-1973	As maximum

Source: "Lighting and Thermal Operations," Federal Energy Administration, 1975.

energy to let a light burn rather than turn it off and on as needed. The amount of energy that switches can save depends on how and what lights the switches control and whether the switches are organized and located so people can use them easily. If different areas of the floor are controlled by separate switches which are then grouped at the door, the chances are that people will switch everything on as they enter the room; therefore each switch should be located close to the area it controls.

One architect designing a school placed three rows of lights parallel to the outside wall in each classroom, with the row nearest the windows on a local switch. This 9-ft strip could easily be lighted by sunlight 75 percent of the time; but in actual practice no one bothered to use, or find, the proper switch. Photocell switching in response to light levels is one answer. Time-clock switches with manual overrides can turn off lights at a certain hour, yet can reactivate them manually as needed. Timer devices can be installed in public bathrooms, corridors, storage and equipment areas, and be set to turn off automatically after occupancy.

Lighting Efficiency

For a given wattage the amount of light will vary with both the kind of lamp used and the fixture in which it is installed. This variation is measured by the coefficient of utilization. As in Table 5-2, efficiencies vary among lamps rated at the same number of watts and among lamps of the same types.

As shown, efficiencies range from about 23 to 180 lumens per watt (lm/W). Most manufacturers have developed lamps specially intended for retrofitting; these use less energy, provide more lumens per watt, are interchangeable with other lamps, provide the user with a choice of wattage, and maintain light distribution and appearance.

Table 5-2 Lamp output
(Lumens per watt)

Source	Efficacy
Incandescent	14–23
Mercury vapor	56–80
Fluorescent	50–85
Metal halide	60–100
High-pressure Sodium	90–130
Low-pressure Sodium	140–180

Where existing lamps are connected in pairs, a "phantom tube" has been developed to permit removal of every other lamp. The phantom tube produces no light, looks like a standard tube, but contains a small capacitor that maintains the electrical balance in the circuit. Tests have shown that pairing a fluorescent tube with a phantom gives the former a life as much as 80 percent greater than normal, and the life of a phantom tube is substantially greater than the 20,000-h utilization expectancy of the standard 48-in tube.

VERTICAL TRANSPORTATION

The amount and kind of vertical transportation are dictated primarily by user needs, and not by energy considerations (which are usually less than 3 percent of a building's total consumption).

Elevators

Elevators in high-rise buildings occupy space that must be considered in the heating, cooling, and lighting load and reduce the ratio of net rentable floor area to gross building area. These elements can be reduced by installing low-rise elevators to serve the lower portion of the building, eliminating these shafts for the upper parts of the building, and having express elevators with no doors in the low-rise section to serve the upper floors.

Elevator shafts cause stack action which increases infiltration and contributes to the heating and cooling load of ventilation air. Shafts should be tightly sealed, elevator doors should open quickly and close tightly, and any mechanical interference with operation should be reduced.

The amount of power required to operate an elevator is a function of the height of the building, the number of elevator stops, the passenger capacity and load factors, and the efficiency of the hoisting mechanism. For example, a 2500-lb-capacity "local" elevator making 150 stops per car-mile consumes about 5 kWh/car-mile, while an "express" elevator making 75 stops per car-mile consumes about 4 kWh/car-mile.

Energy can be saved by using two basic approaches: Eliminate unnecessary use, and completely, or at least partially, shut down elevators and elevator equipment not in use.

The elevator manufacturer is a good source of ideas for saving energy. There are numerous mechanical or electric components which can be modified. Probably the most important is control equipment for the electric motor driving the elevator. An elevator service company or in-house personnel should review the controls to ensure that motors automatically shut off following a 2- to 3-min period of nonuse (known as "putting them to sleep").

Escalators

Escalators are expensive to install and, because they run even when no one is using them, can be expensive to operate. Where a traffic situation makes their use necessary, they should have controls which modify their use during low-traffic periods through use of a treadle switch which operates only on demand. When shut off altogether escalators can still serve as staircases to help move traffic.

OPERATION AND MAINTENANCE

Operation and maintenance is the weakest link in the chain of events that must take place in order to have an energy-efficient building. Shortcomings in the operation of a building can negate much of the effort that went into its planning and construction.

This activity is the function of the building's operators, but the architect should see that mechanical consultants assist this activity by:

- Preparing an operating manual for the structure. The manual should be comprehensive for the complete system and should be suited to the level of comprehension of the user. While these manuals are usually prepared by the consulting mechanical engineer, the architect needs to check that this work is done properly.
- Conducting a meaningful training program for operating personnel and repeating it periodically as personnel change. Participants in this program should include the mechanical engineer, the mechanical subcontractors, and manufacturer's representatives. The operating manuals have to be tailor-made for each system; and the instructions will, of necessity, differ from building to building. Typical of some of the specifics covered in the manuals are sample instructions for boiler maintenance.

Boilers: General

- Inspect boilers for scale deposits, accumulation of sediment, or boiler compounds on waterside surfaces. The rear portion of the boiler must be checked because it is the area most susceptible to the formation of scale. (Scale reduces the efficiency of the boiler and possibly can lead to overheating of furnace, cracking of tube ends, and other problems.)

• Fireside of the furnace and tubes must be inspected for deposits of soot, flyash, and slag. The fireside refractory surface must also be observed. Soot on tubes decreases heat transfer and lowers efficiency. (If the boiler does not now have one, consider installation of a thermometer in the vent outlet. It can save inspection time and often can prove to be more accurate than visual inspection alone.) If the gas outlet temperature rises above normal, it can mean that the tubes need cleaning. Evidence of heavy sooting in short periods could be a signal of too much fuel and not enough air. Adjustment of the air/fuel ratio is required to obtain a clean-burning fire.

• Inspect door gaskets. Replace them if they do not provide a tight seal.

• Keep a daily log of pressure, temperature, and other data obtained from instrumentation. This is the best method available to determine the need for tube and nozzle cleaning, pressure or linkage adjustments, and related measures. Variations from normal can be spotted quickly, enabling immediate action to avoid serious trouble. On an oil-fired unit, indications of problems include an oil pressure drop, which may indicate a plugged strainer, faulty regulating valve, or an air leak in the suction line. An oil temperature drop can indicate temperature control malfunction or a fouled heating element. On a gas-fired unit, a drop in gas pressure can indicate a drop in the gas supply pressure or a malfunctioning regulator.

• Note the firing rate when log entries are made. Realize that even a sharp rise in temperature need not necessarily mean poor combustion or fouled waterside or fireside. During load change, stack temperatures can vary as much as 100°F in 5 min.

• Inspect stacks. They should be free of haze. If not, probably a burner adjustment is necessary.

• Inspect linkages periodically for tightness. Adjust when slippage or jerky movements are observed.

• Observe the fire when the unit shuts down. If the fire does not cut off immediately, it could indicate a faulty solenoid valve. Repair or replace as necessary.

• Inspect the nozzles or cup of oil-fired units on a regular basis. Clean as necessary.

• Check the burner firing period. If it is improper, it could be a sign of faulty controls.

• Check the boiler stack temperature. If it is too high (more than 150°F above steam or water temperature), clean the tubes and adjust the fuel burner.

• Inspect all boiler insulation, refractory, brickwork, and boiler casing for hot spots and air leaks. Repair and seal as necessary.

Boilers: Fuel Oil

- Check and repair oil leaks at pump glands, valves, or relief valves.
- Inspect oil strainers. Replace if dirty.
- Inspect oil heaters to ensure that oil temperatures are being maintained according to manufacturer's or oil supplier's recommendations.

Boilers: Coal-Fired

- Inspect coal-fired stokers, grates, and controls for efficient operation. If ashes contain an excessive amount of unburned coal, it's probably a sign of inefficient operation.

Boilers: Electric

- Inspect electric contacts and working parts of relays and maintain them in good working order.
- Inspect heater elements for cleanliness. Replace as necessary.
- Check controls for proper operation. Adjust as necessary.

Central Furnaces, Makeup Air Heaters, and Unit Heaters

- All heat exchanger surfaces should be kept clean.
- Check air-to-fuel ratio and adjust as necessary.
- Inspect burner couplings and linkages.
- Inspect casing for air leaks and seal as necessary.
- Inspect insulation and repair or replace as necessary.
- Follow guidelines suggested for fan and motor maintenance.

Chapter 6

Recycling and Storing Energy

RECYCLING ENERGY

Recycling means using something more than once. Ideally, energy should not be treated as a disposable. If it can be reused, it makes sense to do so.

As applied to buildings, the term *recycling* relates almost exclusively to thermal energy, whether it be for heating or for cooling. Electricity is almost never capable of being recycled, since when it is used for the first time within a structure, it is converted into either thermal energy or mechanical work. The electricity used to power an elevator is changed into both mechanical motion and heat, and the electricity used to light a lamp is changed into both heat and light. In effect, there is no electricity left to recycle.

But heat and cold are available for reuse if they can be captured before they escape from the envelope enclosing the building.

As described in Chap. 4, thermal energy flows through a building's skin in a constantly changing pattern that varies not only with the nature of the skin itself, but also with what is happening to the temperature both inside and outside the structure at any moment. Heat always flows from a hot body to a cold body, and the process continues until both are at the same temperature and equilibrium is established. In the winter a building wants to heat the great cold outdoors; in summer, the reverse is true.

There are basically two game plans which apply in recycling energy. In the first the energy which is reused would otherwise be wasted, as when we capture the heat which would otherwise go up the chimney and reuse it right away to heat the structure instead of burning more fuel for the same purpose. The second technique is to take heat or cold which has gotten out of balance within the structure and redistribute it, as when we take the heat from an overheated south-facing portion of a structure and send it to the cold north side rather than using fuel to air-condition the south side while simultaneously burning more fuel to heat the north. Implicit in both techniques is the assumption that the energy which is to be recycled will be used immediately.

But supply and demand do not always coincide in time. Sometimes a building generates more heat than it can use during daylight hours on a winter's day, as when an office building's lights and computers produce an excess of heat over and above what is required to keep the building comfortable. At night, with lights and equipment off, that excess heat produced during the day could, if available, be used to heat the building during the cold night.

The key to bridging the time differential is storage, and the techniques for accomplishing it are described in the second half of this chapter. Storage is like a savings bank into which we make (thermal) deposits that can be withdrawn when needed at a later time. This time period can be later on the same day or even later during the year.

We begin by examining ways of recycling energy which is to be reused right away.

INSTANTANEOUS HEAT RECOVERY

The devices used to recover thermal energy fall into two groups: air and liquid. In both cases the objective is to capture the heat before it is rejected from the building.

Air Systems

Heat Wheels (Rotary Heat Exchangers)

As illustrated in Fig. 6-1, this device is in the form of a cylindrical drum or wheel. The wheel rotates in a casing installed between exhaust-air ductwork and outside-air (intake) ductwork. The wheel rotates through both airstreams, absorbing heat from the higher-temperature airstream and transferring it to the cooler airstreams. Where it is desirable to transfer humidity as well as heat, specially designed hygroscopic recovery wheels absorb moisture from the more humid airstream and transfer it in the vapor state to the drier airstream. The wheels work equally well whether they are precooling the incoming air, as in summer operation, or preheating it in winter operation. It is the rotation of the wheel that causes the flow of energy. The exchange surfaces of the wheel are alternately exposed to both airstreams, transferring energy while the two airstreams remain isolated.

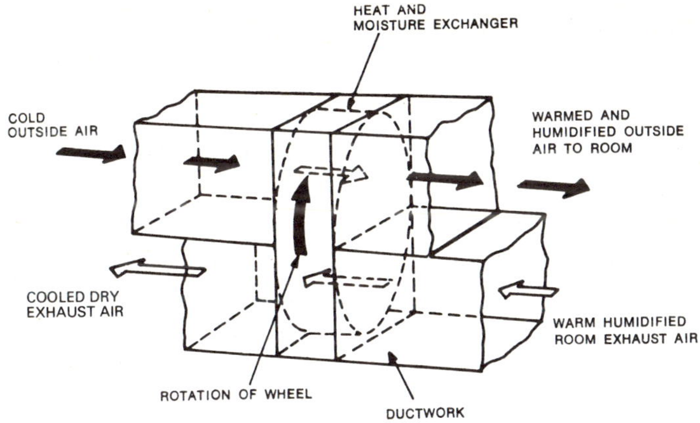

During winter operation, the wheel surface absorbs heat and moisture from the exhaust airstream and transfers it to the fresh-air supply airstream. Up to 90 percent of the normally wasted exhaust energy is recovered to heat and humidify the fresh air supply.

(a)

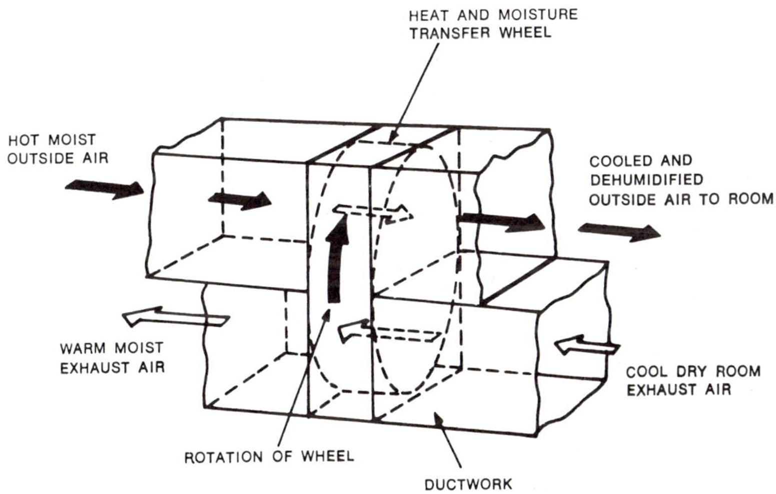

In summer, the wheel surface is cooled and dehumidified as it rotates through the cool, dry exhaust airstream. The surface then rotates to the hot, moist fresh-air supply airstream and absorbs heat and moisture. Up to 90 percent of the normally wasted cooling energy in the exhaust airstream can be recovered to cool and dehumidify the fresh air.

(b)

FIG. 6-1 Rotary heat exchanger. (*a*) Winter operation; (*b*) summer operation. *(The Wing Company.)*

While the supply and exhaust ducts must usually be located near each other in order to use this system, close proximity can occasionally be an obstacle, especially when exhaust streams are contaminated. In such cases, purge sections must be added to the wheel, and special filtration may be necessary in the supply airstream.

Air-to-Air Exchangers

These exchangers transfer heat directly from one airstream to another through direct contact on either side of a metal heat transfer surface. (See Fig. 6-2.) An example of this might be in an institutional kitchen where heat from steam kettles is collected by a hood and is then processed through the air-to-air exchanger to heat the incoming makeup air.

Run-around Coil System

This system is comprised of two or more extended surface fin coils installed in air ducts and interconnected by piping. (See Fig. 6-3.) The coil-type exchanger has the advantage of permitting heat transfer between supply and exhaust systems in widely separated portions of a building, since coils are installed in each of the intake and exhaust ducts and are connected only by piping.

The heat exchanger fluid, usually consisting of ethylene glycol and water, is circulated through the system by a pump, removing heat from the hot airstream and rejecting it into the cold airstream. It is relatively easy to run piping for long distances through a structure, interconnecting a number of separate exhaust and fresh-air systems to form one complete system.

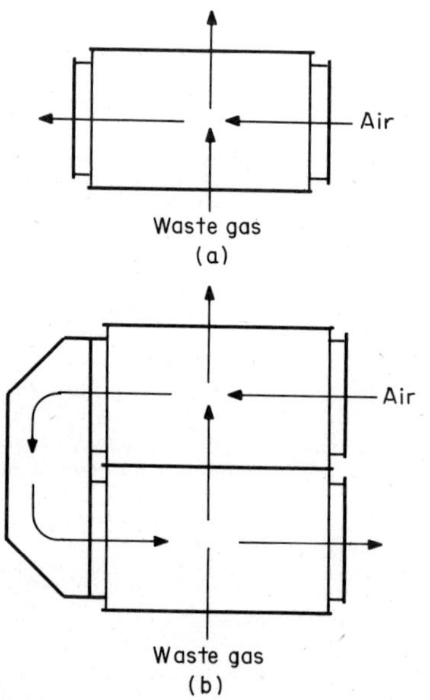

FIG. 6-2 Plate-type heat exchanger. (a) Single-pass recuperator; (b) two-pass recuperator. *(Eclipse Lookout Company.)*

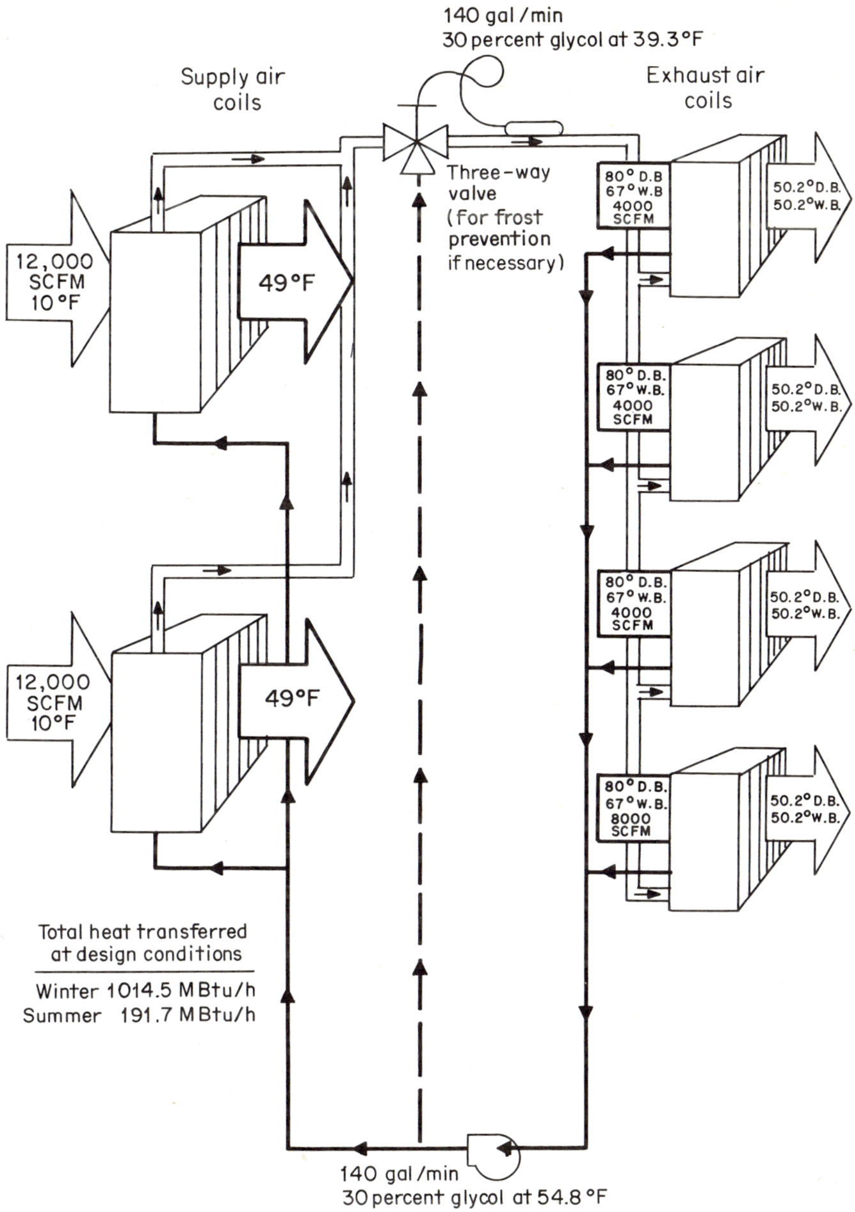

FIG. 6-3 Run-around coil system: Winter operation schematic. *(American Air Filter.)*

Heat Pipe Systems

Heat pipes are composed of extended-surface, finned tubes extending between adjacent air ducts (Fig. 6-4). The tubes run continuously from one duct to the other, and they are on the same horizontal plane. Each tube contains liquid refrigerant, which evaporates at the warm end of the tube, absorbing heat from

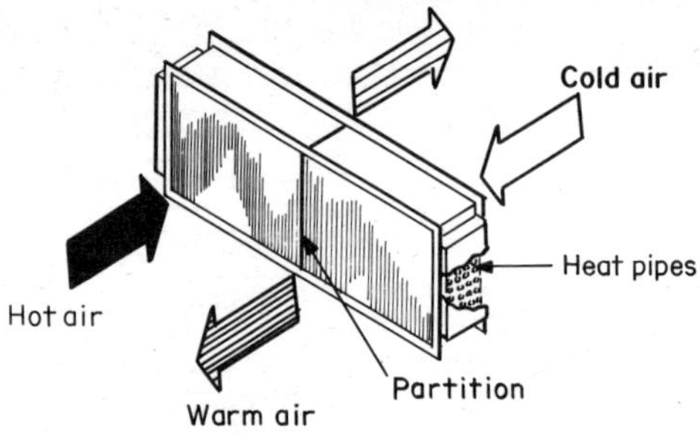

To prevent cross contamination of the airstreams, a sealed partition divides the thermal recovery unit into two sections. The pipes pass through this partition.

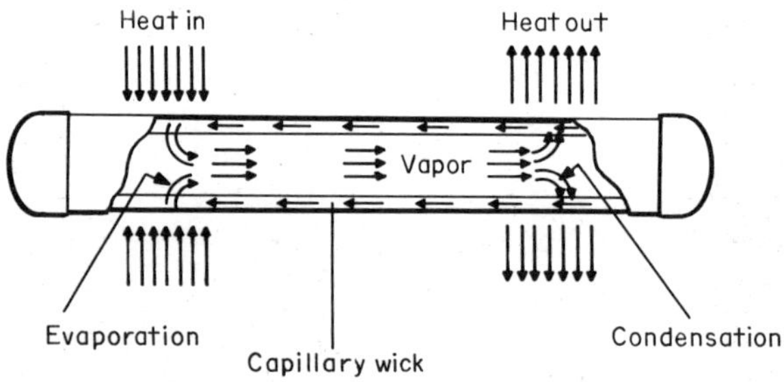

FIG. 6-4 Heat pipe. *(Gaylord Industries Inc.)*

the warm airstream. The evaporated refrigerant then migrates as a gas to the cold end of the tube, where it condenses and releases heat into the cold airstream. The condensed liquid then runs back to the hot end of the tube to complete the cycle. To operate effectively, the heat tubes must slope down from the cold end to the hot end, and accurate control of the slopes must be provided. Heat pipes running between hot and cold air ducts can be used only when the ducts are immediately adjacent to one another, and thus some modification of existing ductwork may be necessary for installation in retrofit situations.

Heat Pumps

These heat pumps are described in Chap. 7. They can transfer energy back and forth between any combination of air or liquid sources.

Liquid Systems

Shell and Tube Heat Exchangers

These heat exchangers can be used to exchange heat from liquid to liquid, steam to liquid, and gas to liquid. All three configurations are sold in a wide range of sizes and outputs, with reliable heat exchange data available to the designer. A particularly favorable application of these exchangers involves the capturing of energy from hot condensate, or from hot refrigerant gas, condenser water, and hot drain lines from kitchens and laundries.

In a laundry, for example, heat can be recovered from laundry wastewater before it is discharged to the sewer. A heat exchanger system uses the 180°F soiled wash water to heat the incoming freshwater from 50 to 130°F. Condensate from the steam-heated iron machines is also used, and it raises the water temperature further from 130 to 170°F. The hot-water heater requires only enough new energy to add 10°F to raise the final water temperature to 180°F. The soiled wash water is finally discharged to the sewer at 72°F (as contrasted with its original 180°F temperature), and the condensate is returned to the power plant at 170°F.

In both air and water exchangers, the devices make it possible to instantaneously reuse energy which might otherwise be exhausted from the structure and thus be wasted. These devices cost money and take up space, however, and it must be emphasized that an architectural/engineering concept that minimizes the amount of air which it is necessary to exhaust is greatly preferred to a scheme which exhausts large volumes and then uses recovery devices to minimize the impact. The ideal combination is minimum exhaust with maximum recovery.

DAILY HEAT RECOVERY

There are many occasions when reclaimed energy cannot be used efficiently at the moment of capture but is needed at some other time of the day. A house located in the New Mexico desert, for example, requires cooling during the day but must be heated at night to compensate for daytime temperatures above 100°F and nighttime temperatures below 50°F. Heat extracted during the daytime cooling process has no immediate use (except for the possible heating of domestic hot water). The extracted heat would be useful at night to counteract the desert chill. In order to effect this time delay, the heat must be stored until needed and then released at the appropriate moment.

This delay is accomplished by storing the heat in one of two ways. The building structure itself can act as a thermal sink to store the heat until needed, or the heat can be deposited in special reservoirs constructed within or adjacent to the structure. One way in which a house can be designed to act as this thermal flywheel is illustrated by David Wright's sundown house in Sea Ranch, California. (Fig. 6-5).

The house is set in a picturesque northern California coastal meadow and is an example of a direct-gain passive solar design. (For other examples of passive design see Chap. 4.) In a direct-gain structure, thermal solar energy moves from sun to living space to storage mass. Solar radiation is collected in the living space and then stored in a thermal storage mass. Thus, the actual living space is directly heated by the sun and serves as a live-in collector. In turn, this direct

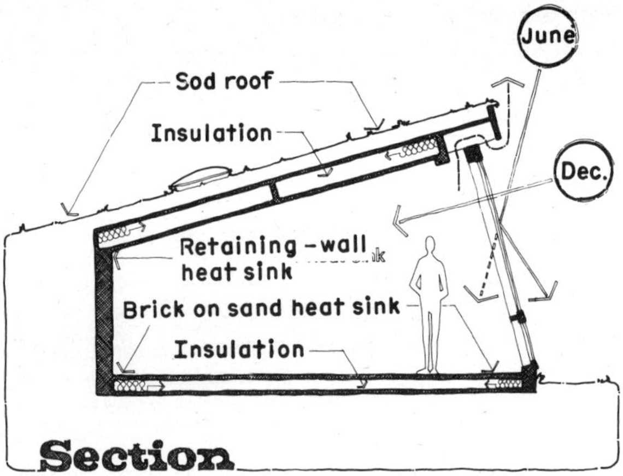

FIG. 6-5 David Wright's sundown house. *(AIA Journal, April 1978.)*

heat gain is absorbed and stored in the massive walls of dense concrete and floor of brick or sand for nighttime reradiation.

The ability to store heat in water, concrete, rocks, etc., is a function of the mass (weight) of the storing material, its specific heat (in British thermal units per pound per degree Fahrenheit), and the temperature increase of the material. The specific heat of water is 1.0 Btu/(lb·°F), while that of rocks is 0.2 Btu/(lb·°F). In order to store a given amount of heat, therefore (assuming the same increase in temperature for both), 5 times the weight of rocks will be needed as compared to water.

In the case of both rocks and water, storage is achieved by making the material hotter. Another way of storing heat is to use it to melt certain chemicals. Per pound, these materials (known as phase-change, or heat-of-fusion, materials) can absorb approximately 100 times the heat that 1 lb of water can, or 500 times that of rock. It should be noted that this storage is accomplished with no increase in the temperature of the storing material, because the thermal energy is used to change the state of the material from solid to liquid. When heat is withdrawn from chemical storage, the material changes back to a solid state. Chemical storage thus has the ability to hold large amounts of heat in a small volume and weight, while at the same time accomplishing this at no change in temperature (thus substantially reducing insulation requirements of the storage container).

An illustration of the use of chemical storage is the Massachusetts Institute of Technology's Solar Building 5. In this structure polymer concrete ceiling tiles, 2 ft square and 1 in thick, are the storage component of the system (Fig. 6-6).

The tiles have a chemical core which stores a day's heat and then releases the retained heat as needed. Because the core stores heat at 73°F, it maintains a nearly constant room temperature and thus prevents overheating, which normally is wasted heat. At night, as the outside temperatures lower, the chemical core sealed within the tile parcels out enough heat to maintain room temperature near its daytime level. The core, therefore, acts as a built-in thermostat to stabilize the temperature in the room. The tile stores 220 Btu/ft^2 and weighs 11 lb/ft^2.

The previous examples illustrate some of the ways in which the building structure itself can be used to store thermal energy. When storage consists of separate tanks, either within or outside the building, it can be used not only to store reclaimed heat but also to take advantage of cheap off-peak electric rates available in certain areas.

If a sealed tank is used, as shown in Fig. 6-7, then greater pressure can also be used in storing the water; in turn, this permits higher water temperatures to be used for storage, sometimes as high as 300°F. This higher temperature allows more heat to be stored in a given volume. When used to take advantage of off-peak rates, the thermal storage system looks like an ordinary hot-water heating tank. In operation, water is heated by electric immersion heaters at the bottom of the storage tank. Inserted below the waterline are the primary and secondary heat exchangers, which can supply water at the specific temperature required by the system.

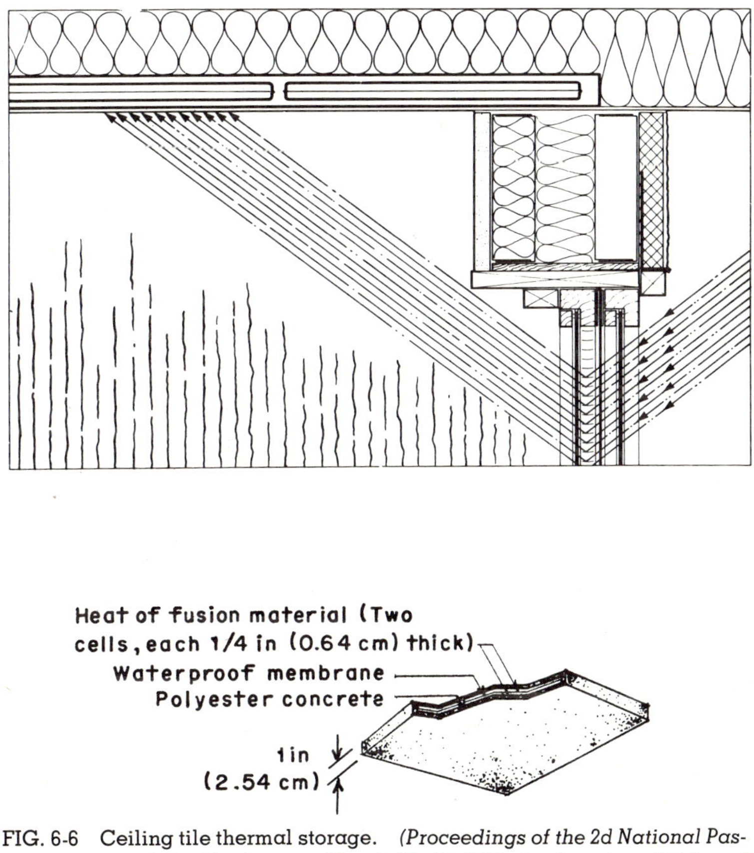

FIG. 6-6 Ceiling tile thermal storage. *(Proceedings of the 2d National Passive Solar Conference, March 16–18, 1978; the American Section of the International Solar Engineering Society [ISES], vol. 2.2.)*

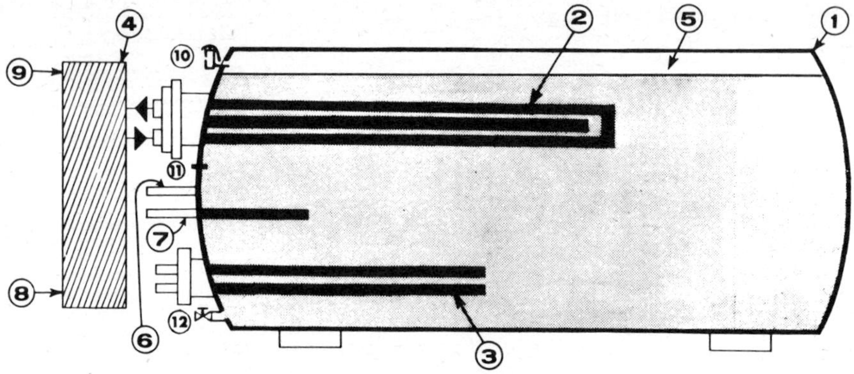

FIG. 6-7 Pressurized water storage system. The basic megatherm unit: 1. Standard steel tank; 2. heat exchanger; 3. electrical elements; 4. tempering condenser; 5. expansion chamber; 6. pressure control; 7. temperature safety controller; 8. inlet; 9. outlet; 10. relief valve; 11. low-water cutoff; 12. drain. *(Megatherm Corporation.)*

ANNUAL HEAT RECOVERY

The concept of annual heat recovery is an intriguing one. At present we use energy to heat buildings in winter and then use additional energy to cool them in summer. In effect, our fuel bills are the total of our winter costs plus our summer costs. If it were possible to take the heat we reject in summer during the cooling phase and reuse it the following winter during the heating period, then our annual fuel bill would be greatly reduced.

In order to accomplish this annual exchange, we need enough storage to hold a season's heat. One way of accomplishing this is illustrated by the Annual Cycle Energy System (ACES) house. This system provides heat from the latent heat of fusion of water and uses the ice frozen in the winter to provide air conditioning in the summer.

The annual-cycle energy system works in the following manner:

1 Heat is obtained by freezing water to remove the heat of fusion (144 Btu/lb). The heat thus obtained is pumped by a one-direction heat pump to deliver heat to the building.

2 The ice frozen during the heating season is stored and used to provide air conditioning during the summertime. This stores the summertime solar energy in the liquid water by increasing its enthalpy as it passes from solid to liquid state.

The cycle is repeated each year, and the energy cost is the cost of operating the heat pump during the heating season. The heat pump is designed to operate between the fixed limits of 22°F evaporating to 105°F condensing; and since the system is optimized for these temperatures, it is expected that the coefficient of performance (COP) on the heating cycle will exceed 3.5. The overall yearly COP is between 5.0 and 6.0.

The ACES concept dealt with annual storage for a single house. On a larger scale the possibility exists of providing thermal storage on a communitywide basis. One such suggestion currently being investigated involves the use of large underground bodies of water (aquifers) which have sufficient capacity to serve the communities located above or near them.

Dr. W. Richard Powell of Johns Hopkins University has proposed an aquifer system of storing summer heat and winter cold for use in the opposite seasons which might reduce the energy consumed for heating and cooling by as much as 80 percent. In effect, the aquifer acts as a thermal utility to provide heating and cooling service to a coummunity via warm- and cold-water pipelines from the central storage. In winter, heat pumps in the buildings would extract heat from 60°F water which had been pumped into a warm-water storage aquifer the previous summer when warm water was plentiful. Warming the community via heat pumps in winter, the summer water would be reduced in temperature to about 40°F and returned as cold water to its own aquifer to be stored for economical airconditioning the following summer.

In summer, the cold water drawn from the cold-water aquifer and distributed through the community pipeline network to cooling coils, which serve as the airconditioning units, would in turn be warmed to 60°F for aquifer storage for the following winter.

With the proposed community system, a homeowner could rent heating and cooling units as she or he rents a telephone. Without a need for furnace, flue pipe, or air conditioner, the price of the home could be reduced.

In both the ACES and the Powell aquifer concepts, storage is accomplished by heating or cooling a body of water. Earlier in this chapter we discussed the storage possibilities of using phase-change chemicals for compact storage.

Of even greater interest, albeit a long way from commercial availability, are the possibilities involved in reversible chemical reactions. These reactions can be stated as

$$\text{Chemical A} + \text{heat} \rightleftarrows \text{Chemical B} + \text{water}$$

The left side of the equation states that if we have a chemical and heat it (as with heat withdrawn from a building during the cooling season), then the original chemical will change into another chemical while releasing water vapor. Since the reaction is reversible, when we add water to the new chemical, it will return to its original state while giving off heat which can be used to heat the structure. Until the water is added, however, chemical B is perfectly stable and can be stored at room temperature. Because the energy that 1 lb of reversible material can hold is many times greater than even that of phase-change materials, it promises great compactness.

In effect, the addition of heat to chemical A creates a synthetic fuel, chemical B, which can be made to release its heat by adding water to it. Unfortunately, most of the reversible reactions take place at temperatures over 800°F, but the search is continuing for lower-temperature materials. Should these efforts prove

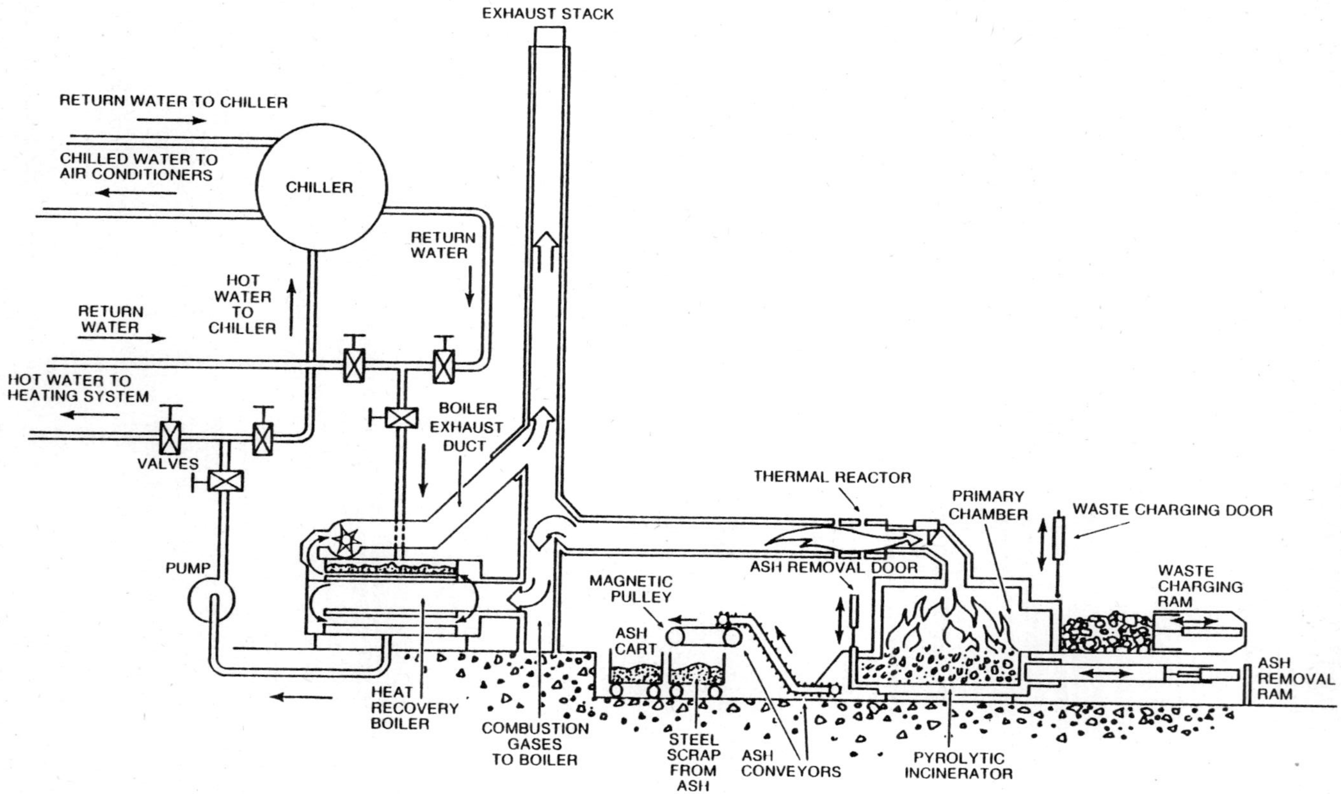

FIG. 6-8 Combustion of industrial and agricultural waste. *(Rockwell International.)*

successful, the possibility of self-contained room units which could both heat and cool without any outside fuel needs might become a reality. This assumes that enough heat can be extracted from cooling to supply that needed for heating, and that somehow the energy needed to operate the equipment can also be obtained from the thermal energy of the space.

INDUSTRIAL-WASTE FUELS

Many industrial plants generate large amounts of waste which in the past had to be carted away to landfills or other disposal and which created environmental burdens for the community and expense to the industry paying for the removal.

Under the impact of the energy crunch, many of these waste products are being reexamined for their fuel capability. Wood scraps, sawdust, hogged wood and bark, and paper are but a few examples of material that can be burned in-house to provide a cheap source of energy. Because the available waste may not always coincide with the needs of the plant, the available waste may have to be supplemented by conventional fuels. Specially designed boilers must therefore be used which can accommodate both conventional fuel and the industrial wastes.

In its Marysville, Ohio, plant Rockwell International uses large volumes (1500 tons) of combustible waste that were previously hauled to a landfill. In addition to the above waste, the company has 50 acres of unused land planted in corn. The farmer gets the corn crop and Rockwell gets approximately 150 tons of corn stalks, cobs, and so on, for fuel. Rockwell estimates that through use of this waste, their annual gas bill has been cut in half.

The system is comprised of several major components, the first of which is a pyrolitic incinerator (see Fig. 6-8). The stack gases (or heat) from the waste are drawn through a specially constructed waste-heat boiler (the second component) which operates at a higher rate of efficiency than a conventional boiler of the same British thermal unit rating.

In lieu of a burner, the boiler has a forced-draft fan which draws the hot stack gases through the boiler when heat is needed; and when demand is satisfied, the fan simply shuts down, allowing the gases to go up the stack bypassing the boiler. In winter months the hot water is circulated to nine rooftop heating and air-conditioning units and used to heat the plant. In summer months, use is made of the third component which is an absorption chiller. This unit uses hot water from the boiler to vaporize the lithium bromide solution of the absorption chiller.

Chapter 7

Emerging Energy Conservation Equipment

Since the 1973–1974 oil crisis, the marketplace has witnessed the introduction of a large variety of energy conservation devices and equipment. These represent both the efforts of private industry to cash in on the fuel shortage and the efforts of government-sponsored research to find effective ways of conserving energy.

As with all new technologies, some of the new ideas will catch on, while other will drop by the wayside. Many of the new devices will have bugs in them which only experience will reveal. The sampling that follows therefore is intended to indicate what some people involved in the industry have been thinking about. As time passes, newer equipment will be created and replace some of the items shown below.

Material has been selected which will be of particular interest to architects. Similar, but more complex, innovations are occurring in all phases of the electrical and mechanical engineering fields. Discussion with consultants will acquaint the architect with many of these ideas.

Electrocomp Energy Computer

This portable computer (Fig. 7-1) is designed to calculate heat loss and annual heating cost of residences. After feeding in the appropriate building information together with climatic references, a meter immediately indicates the heat loss in British thermal unit-hours (Btu·h) or the annual operating cost in dollars.

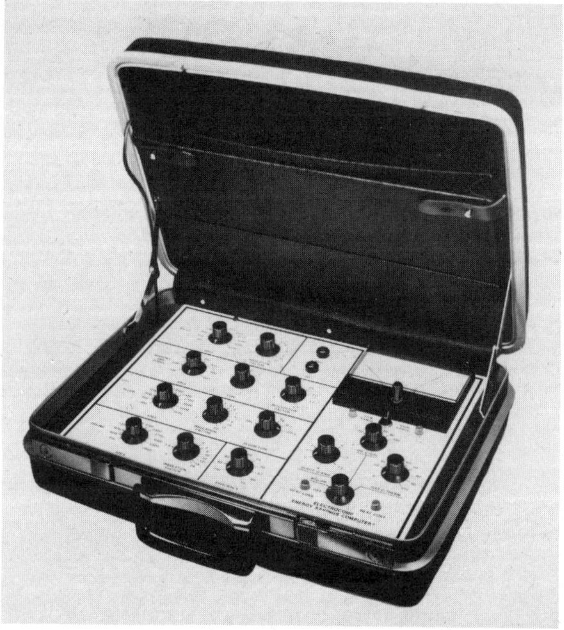

FIG. 7-1 Electrocomp energy computer. *(Beckman Instruments, Inc.)*

Flexible Strip Door

Built of vertical plastic strips, these doors provide a thermal barrier at exterior openings (such as loading docks) or between interior spaces having different temperatures (Fig. 7-2). They are designed to permit truck and foot traffic to pass through and are transparent to provide the visibility needed for traffic safety.

Air Doors

These "doors" (Fig. 7-3) provide an air barrier which acts as a thermal barrier to protect both exterior and interior openings against heat transfer. With the heater fan on, the doors act as protection against cold, wintry drafts entering indoor heated areas. Without the heater and with only the fan on, the doors are used in warm weather to prevent loss of air conditioning and provide an effective barrier against insects and dust.

Air-Jet Makeup Air System

Designed to capture the stratified heat which pockets in the upper regions of industrial-type buildings, this system feeds the makeup air through a large plastic tube with punched holes along both sides (Fig. 7-4). The makeup air, as it jets through these holes, aerodynamically mixes with the stratified upper air whose heat tempers the makeup air as it travels to the worker occupancy level.

FIG. 7-2 Dyna flexible strip doors. *(Dynaforce Corporation.)*

Thermatrol Windows

Designed primarily for industrial, commercial, and institutional uses, this inside-mounted thermal-barrier glazing system is made of plastic. It is custom-fitted to each opening and available in bronze solar tint, and panels can be removed for cleaning and ventilation. It is cheaper than the traditional exterior-mounted glass storm sash.

FIG. 7-3 Mars air door. *(Mars Sales.)*

FIG. 7-4 Air-jet makeup air system. *(Dynaforce Corporation.)*

Add-a-Sash Window

Similar to the traditional storm sash, and with an aluminum frame and glass glazing, this system is designed for installation on the inside rather than the outside of the building window. It is available in both double-hung and horizontal slider versions.

Thermal-Barrier Windows

Double or triple glazing has traditionally provided increased thermal resistance through glazed portions of the windows, but the metal sash has provided a conductive end run for thermal transmission. The thermal-barrier window stops this transmission by providing a complete thermal separation between the metal of the interior and exterior frames. It is available for both new and retrofit installations.

Retrofit Double Glazing

Designed for commercial and institutional applications, this method adds an additional pane of glass while retaining the original glazing and sash. The second pane is hermetically sealed to the original glass at their perimeters, forming an insulation space of dead air between the two panes.

Polyvinyl Chloride (PVC) Windows

This plastic sash is available in different window types and with different types of glazing (Fig. 7-5). The sash needs no thermal barrier because, unlike metal windows, the frame itself is a thermal barrier; and because the sash elements are fused together during manufacture, infiltration is significantly reduced.

FIG. 7-5 PVC windows. *(Trocal Window Systems/ Dynamit Nobel.)*

Window Films

Designed to retrofit existing clear-glass installations, these films are designed to reduce the impact of solar energy on the cooling system (Fig. 7-6). They are usually made from a polyester sheet film which is coated with a reflective metal in a vacuum chamber. A pressure-sensitive adhesive bonds the film to the inside of the existing glazing.

Solar Screens

Designed for installation outside existing windows, these fiber glass screens block out 65 to 75 percent of the sun's heat and glare before it reaches the window glass (Fig. 7-7). Designed primarily to reduce cooling costs, the screens can be unlatched and stored during the winter.

Solar Pool Blanket

Much of the need for heating swimming pools derives from the cooling effect of water evaporating from the surface of the pool. This plastic blanket unreels during nonuse periods to prevent this evaporation while at the same time permitting the solar energy from the sun to pass through and be absorbed by the water. In effect, it turns a swimming pool into a large solar collector.

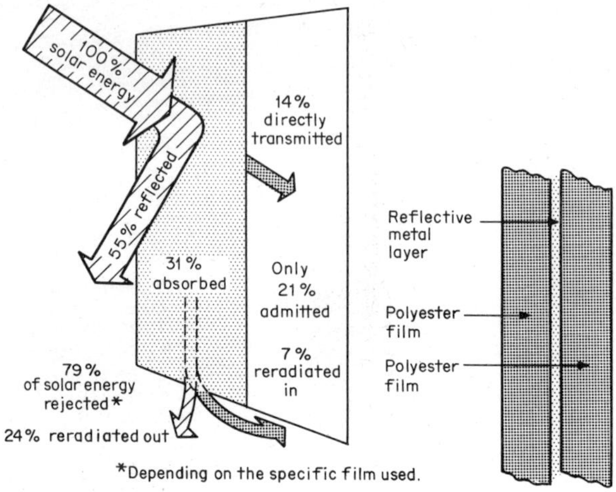

FIG. 7-6 Reflecto-shield. *(MADICO.)*

Foam Sealants

While wall insulation reduces conduction transmission, it generally does not protect against air infiltration under sill plates, around window and door frames, through cracks in corner framing, around electrical boxes, and around hose bibbs. Foam sealants are designed to penetrate these locations and prevent air infiltration.

FIG. 7-7 Solar screen. *(PPG Industries, Inc.)*

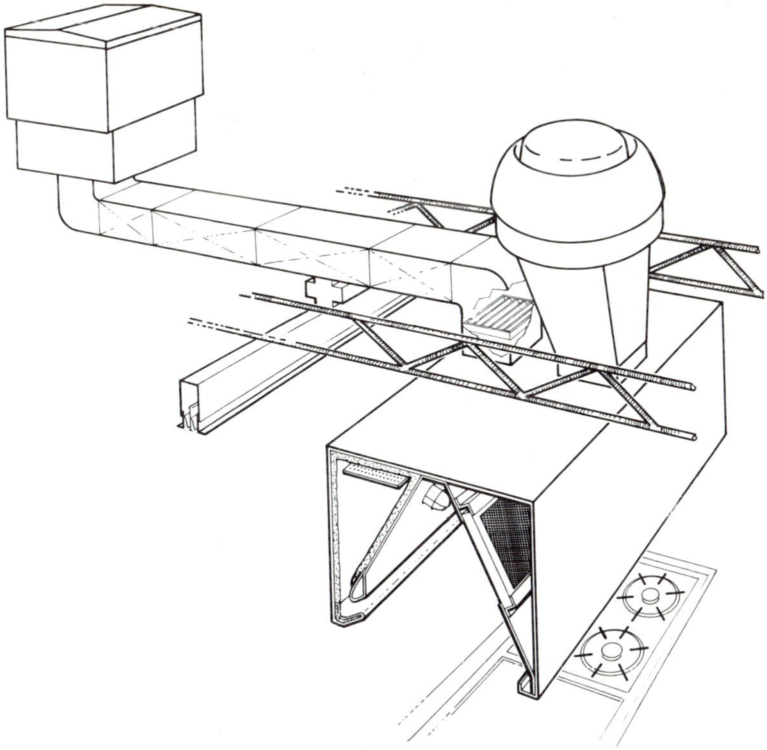

FIG. 7-8 Commercial kitchen exhausts. *(Econovent Systems, Inc.)*

Commercial Kitchen Exhausts

Conventional range-hood exhaust systems waste energy by exhausting 100 percent tempered air from the kitchen area (Fig. 7-8). New exhaust systems end this waste by supplying free *un*tempered air directly to the hood area which is immediately exhausted by the hood exhaust as it removes grease, vapors, and smoke from the capture area.

Thermostatic Attic Ventilators

Designed to reduce cooling loads in residences, the thermostat of this attic exhaust fan turns it on at 100°F and off at 85° F (Fig. 7-9). The thermally protected motor shuts off automatically in the event of fire or extreme temperature.

Hydroplace and Hydrohearth

Unlike most fireplaces, which heat only the room where they are located, these units are designed to take the heat from the fireplace and put it into the central heating system for distribution throughout the house (Fig. 7-10). Water circulates throughout the unit's hollow walls and grates. This water absorbs heat from the fire and transfers it to the central system. It does this directly in the case of a hot-water system and via a heat exchanger in the case of an air system.

FIG. 7-9 Attic power vent. *(Lomanco, Inc.)*

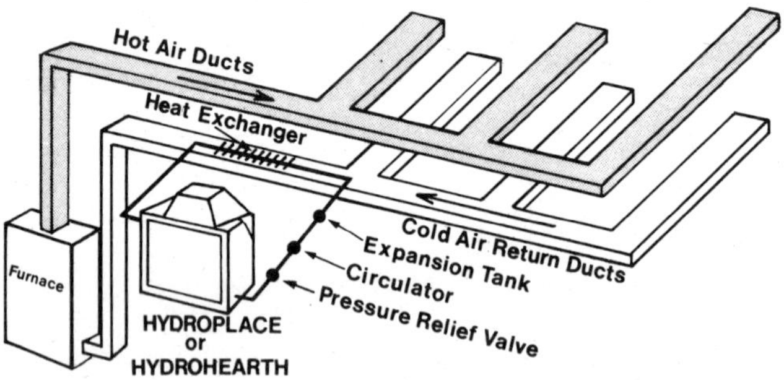

(a)

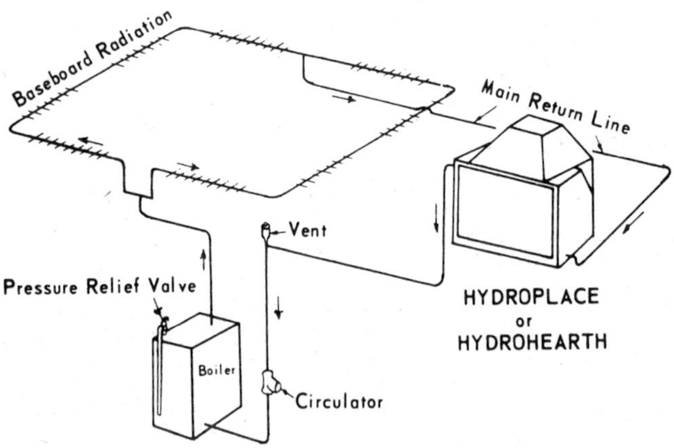

(b)

FIG. 7-10 Hydroplace and hydrohearth. (*a*) Forced-air or heat pump system; (*b*) hot-water system. *(Hydroheat Sales & Distributing Corp.)*

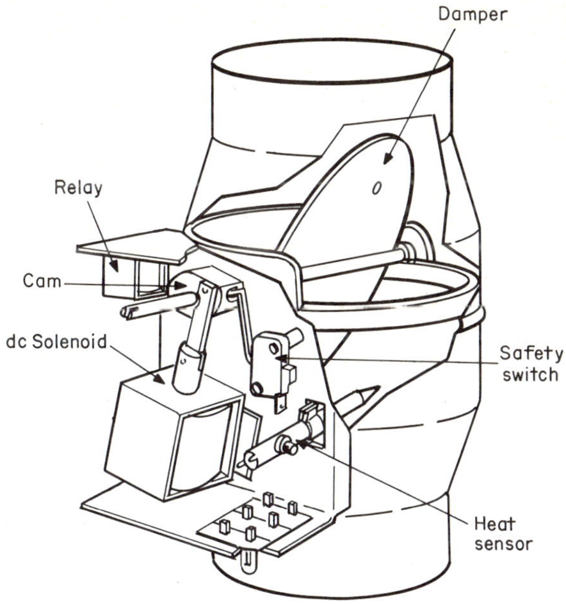

FIG. 7-11 Flue damper. *(Energy Vent, Inc.)*

Flue Damper

This device is designed to keep heat from escaping up the flue during the off cycle of the furnace burner (Fig. 7-11). It automatically puts the damper in the closed position when the thermostat shuts off the burner. A number of safety interlocks are provided.

Flue Heat Exchanger

This air-to-air heat exchanger is designed to be installed in a furnace smokepipe to capture the heat that would otherwise go up the flue (Fig. 7-12). The heated discharge air can then be ducted to another space.

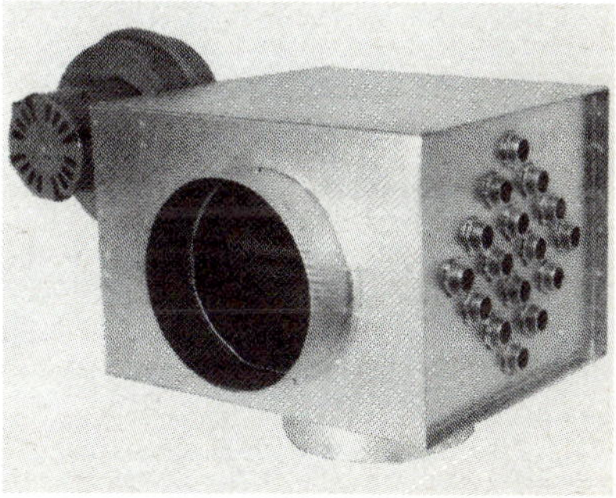

FIG. 7-12 Flue heat exchanger. *(The Supply Company.)*

Computerized Hot-Water System Control

Instead of maintaining the same constant high boiler temperature regardless of weather conditions, as in a conventional system, this control continuously monitors and adjusts the water temperature in the system based on the prevailing outside temperature and the thermostat demand. The hot-water circulator operates constantly and thus continuously picks up heat from the boiler which would otherwise be dissipated during off periods.

Tamperproof Thermostat

Designed to keep control of temperatures in the hands of the building operator rather than the tenant, this thermostat substitutes preset sensors for the traditional bimetallic coil (Fig. 7-13). The sensors are preset for a given temperature, snapped inside the box, and covered with a tamperproof locked cover. A limiter model is also available with preset upper and lower temperature limits; within this range, the tenant can still operate the existing thermostat.

Light-Sensing Thermostat

In addition to its heat-sensing elements, this thermostat constantly monitors lighting levels in the space in which it is installed. When lights are off or low, the thermostat automatically sets back heating and air conditioning to off-occupancy levels.

Terminal Unit Thermostats

When, for whatever reason, some spaces in a building are too hot or too cold with consequent waste of energy, these thermostats are designed to be installed directly on radiators, convectors, or baseboards. They sense the room temperature and control the flow of steam or hot water through a valve that opens or closes in response to the room thermostat.

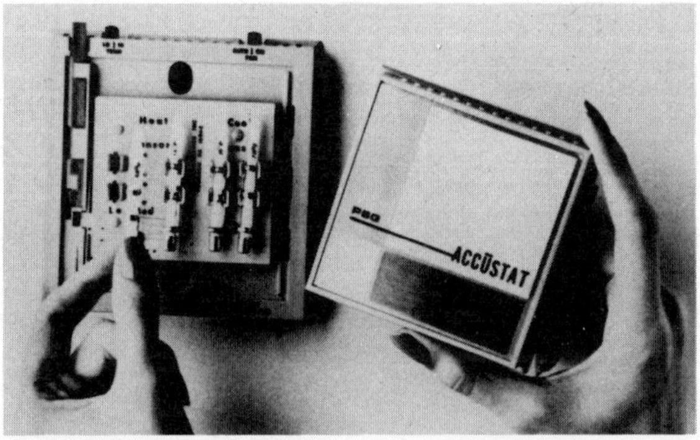

FIG. 7-13 Tamperproof thermostat. *(PSG Industries, Inc.)*

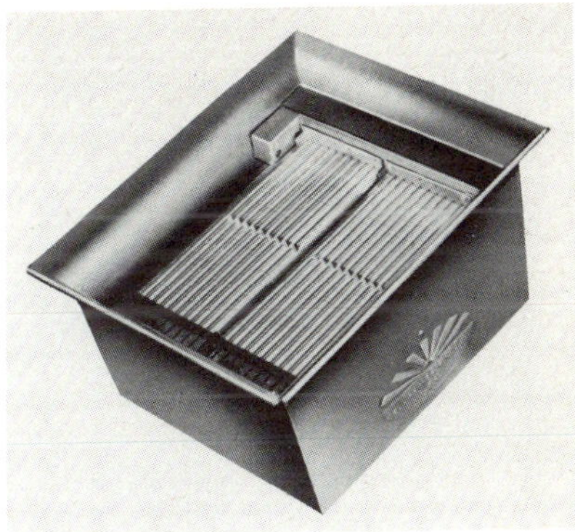

FIG. 7-14 Infrared gas heater. *(Detroit Radiant Products Company.)*

Infrared Gas Heaters

Conventional gas-fired unit heaters heat the air which, in high bay industrial plants, pockets at the ceiling (Fig. 7-14). Infrared heaters get their effect largely through radiant energy and thus are felt by the occupant at floor level who is kept comfortable in spite of much lower ambient temperatures. Applications include not only manufacturing spaces but also such semioutdoor locations as loading docks.

Booster Fans

In central heating systems, air is distributed through ducts, and its volume is controlled by dampers and shutters (Fig. 7-15). These restrictions require blowers to operate against high pressure. The booster fan, mounted into ducts, replaces the dampers and shutters and provides continuous air circulation so that all the furnace heat is captured and little is lost up the chimney.

Ceiling Fans

In commercial/industrial structures where ceiling heights exceed 12 ft, the heat that is stratified toward the ceiling can be recirculated downward by installing

FIG. 7-15 Air duct booster fan. *(Dale and Associates Inc.)*

FIG. 7-16 Heat reclaim fan. *(Emerson-Chromalox.)*

these "ice cream parlor" fans (Fig. 7-16). Depending on fan diameter and ceiling height, each fan can handle between 1000 and 3000 ft^2 of floor area.

Heat Inverter

Another approach to the problem of heat stratification, this unit consists of a duct that incorporates a fan to suck the heat from the upper level and redistribute it back down to the lower level. The ducts come in 8-ft lengths and can be joined to reach any ceiling height.

Energy Monitors (Electric)

These monitors come in several versions. One is designed to measure electric energy consumption on branch circuits and thus aid in energy audits (Fig. 7-17). Another shows the costs, in cents per hour, of the electricity being used in

FIG. 7-17 Electric energy monitor. *(Franklin Electric.)*

the structure at any given time. Usually placed next to the thermostat, it is intended to act as a psychological deterrent to wasteful usage.

Energy Management Systems

These systems come in a large variety of forms to correspond to different building needs. This system is designed for installation in medium-size facilities such as supermarkets, schools, motels, hotels, and office buildings. The control system reduces electrical bills by limiting the kilowatt demand peak and by reducing the kilowatthour energy and fuel adjustment charges. These two functions are known as demand control and load cycling.

Fuel Cells

The fuel cell is an electric power generator which is of particular interest to architects because it can provide all a building's energy needs—electrical, heating, and air conditioning—while often using less fuel than conventional heating needs alone (Fig. 7-18). Unlike traditional power plants, the fuel cell generates electricity through a chemical reaction in a process that is efficient, clean, and quiet. The cells can be located in individual buildings or in urban locations and bypass the environmental concerns associated with nuclear power generation. The technology for this technique is still being developed but has moved from the laboratory to the demonstration stage and should be available in the marketplace in the near future. Fuel cells are being combined with

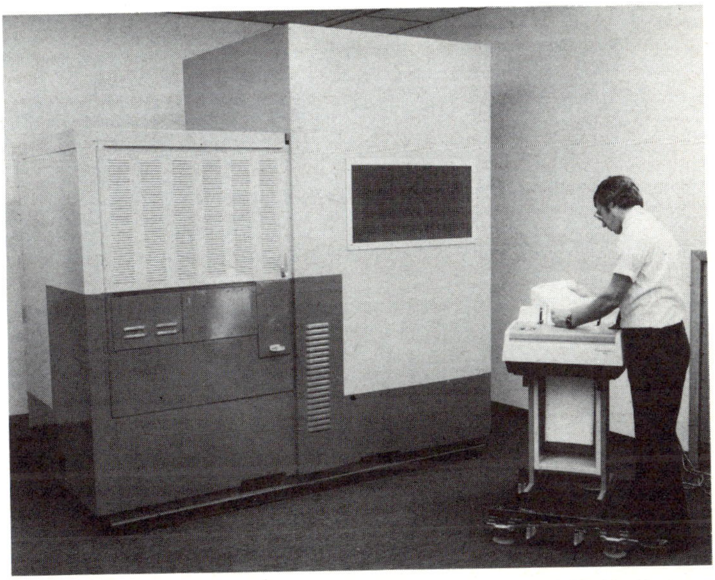

FIG. 7-18 A 40-kW-fuel-cell pilot power plant. *(United Technologies Corporation.)*

heat pump systems, in which heat recovered from the fuel cell is used to drive the heat pump.

Fluorescent Retrofitting

Designed to capitalize on the higher efficiencies enjoyed by fluorescent lighting as compared with incandescent, a series of lighting fixtures is designed to make the substitution easier. Many of the fixtures are designed to simply screw into the old incandescent sockets.

Phantom Fluorescent Tubes

In retrofitting, it is often desirable to cut out every other fluorescent tube. Unfortunately most fluorescent lamp circuits are of the two-lamp series-sequence type in which removal of one lamp causes the other to go out or burn dimly. When phantom tubes replace the tube to be cut out, the second lamp remains at full brilliancy (Fig. 7-19). The phantom consumes negligible power, and power savings are proportional to the number of lamps replaced.

Reducer Fluorescents

Where it is desirable to reduce the amount of light, but to do so without risking the spotty effect of removing tubes, reducer fluorescents can be substituted for existing tubes. They provide energy savings of 33 to 50 percent, while reducing light output correspondingly.

Task Lighting

The switch from overall ambient lighting to task lighting involves considerations of ceiling height, wall color, number of people in the area, type of work being done, and number of windows (Fig. 7-20). Task lighting works particularly well

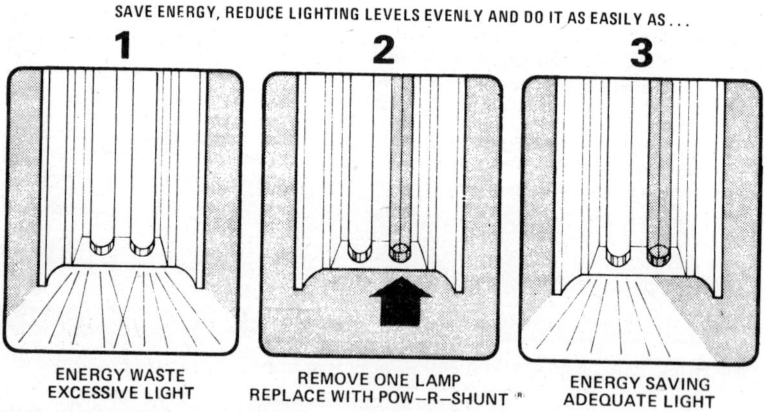

FIG. 7-19 Enersave Pow-R-Shunt. *(Enersave, Inc.)*

FIG. 7-20 UPS/open-plan system with task/ambient illumination. *(JG Furniture.)*

with an open floor plan, and manufacturers of office furniture are marketing increasing numbers of task-lighting work stations.

Demand-Controlled Lighting

Many rooms, especially in office buildings, receive a substantial amount of sunlight during the day. Energy costs can be reduced by continuously adjusting the artificial lighting level to changes in sunlight. The adjustments should be automatic since people rarely can be counted on to do the necessary switching. A photocell and switch can meet this need when used as part of a loop-wiring plan. Infrared motion (people) detectors can turn off lights when people have left.

Plumbing Water Controllers

Limiting the amount of water used in a building produces two types of savings: a reduction in water use charges and a reduction in the fuel needed to heat hot water. New water closets are designed to operate with 3 to 3.5 gal per flush as compared with previous consumption of over 6 gal (Fig. 7-21).

Low-Volume Showerheads and Sink Faucets

Conventional shower heads consume between 6.5 and 15 gal/ min; the low-volume shower heads use between 2.5 and 3.5 gal/min with corresponding savings in hot-water consumption. Conventional faucets use between 6 and 8 gal/min as compared to the new low-volume type which uses about 2 gal/min.

FIG. 7-21 Watersaver closet. *(Eljer Plumbingware.)*

Water Pressure Regulators

Another approach to water conservation, these regulators keep supply water pressure from exceeding preset limits regardless of changes in street pressure (Fig. 7-22). Since the volume of water used depends not only on the plumbing design but also on the pressure at which water is introduced, these regulators provide another control to limit the amount of water consumed.

Ultrasonic Infiltration Detector

Infiltration in a completed structure is often hard to detect. One suggested approach consists of placing an ultrasonic tone generator inside the room to be tested for leaks. The output will not pass through solid material, but will flow

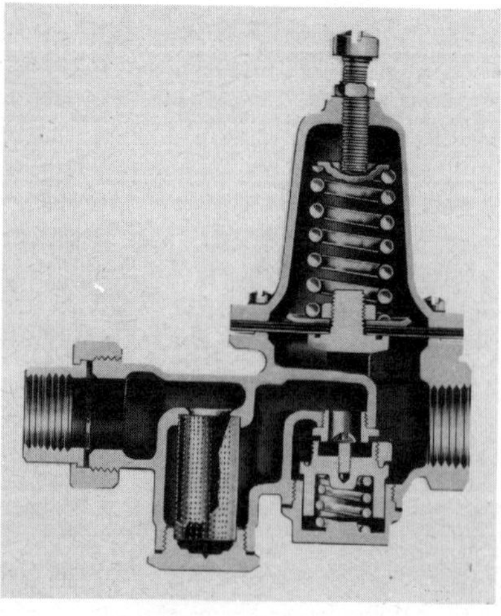

FIG. 7-22 Water pressure regulator. *(Watts Regulator Company.)*

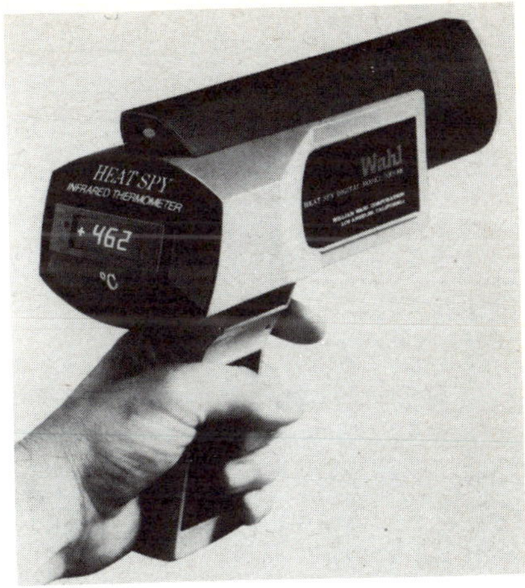

FIG. 7-23 Heat spy infrared thermometer. *(William Wahl Corporation.)*

through rather small openings. This passage can be spotted from the outside by a detector sensitive to the ultrasonic tone and thus can be used to locate the point of leakage.

Heat Detectors

Conventional thermographic equipment (see Chap. 8) is a very useful tool for heat loss detection. The equipment is very expensive (costing more than $50,-000), and a number of less expensive, albeit not as comprehensive, substitutes are coming on the market. One device (Fig. 7-23) which looks like a pistol continuously provides a digital printout of the temperature of the surface at which it is pointed. Another (Fig. 7-24), which looks like a press camera, is pointed at the structure and shows where hot spots or cold spots are located.

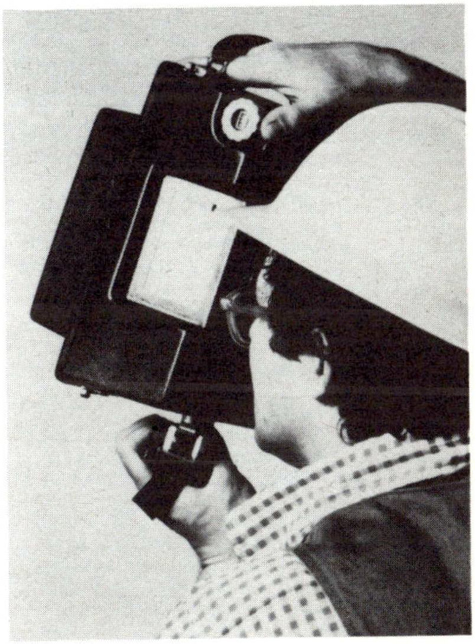

FIG. 7-24 ThermAtrace thermal line scanner. *(Barnes Engineering Company.)*

Heat Pumps

These pumps are described in standard mechanical engineering texts. They work like a refrigerator or air conditioner; they transfer the heat present in one place at a relatively low temperature to another place at a higher temperature. They are reversible and can be used for either heating or cooling.

With the uncertain future of oil and natural gas supplies, the heat pump offers an important opportunity for energy conservation. There are, however, practical problems in comparing the use of heat pumps with electric resistance systems.

Since the heat pump provides both heating and cooling, the decision to use a heat pump will, in most cases, depend on whether central air conditioning and a ducting system are also needed. If summer air conditioning is not needed, it is likely that the heat pump's higher first cost will make it uneconomical on a life-cycle basis.

At present heat pumps are rated under steady-state, full-load operating conditions and are given an energy efficiency ratio (EER) value. In practical operation as a heating device, the efficiency of a heat pump is reduced by intermittent operation, frost buildup, and the process of defrosting. Because of these losses, the actual seasonal performance of a heat pump will usually be less than the predicted seasonal performance which would be calculated from a manufacturer's performance data. The size of the discrepancy is likely to vary among products and will depend on the climate in which the heat pump operates. A recent study at the National Bureau of Standards indicated that these factors resulted in an almost 20 percent drop in the seasonal performance factor (SPF) on a typical residential heat pump operated in the Washington, D.C., area.

Chapter 8

Retrofitting Existing Buildings: Data Gathering

Opportunities to reduce energy use in existing buildings differ markedly from those in the design of new buildings. Most of the major determinants in the physical apparatus have been established. These include the building configuration and orientation, its materials and construction, its mechanical systems and controls, and its specific location. Most of these are difficult and expensive to modify. Some are impossible. On the other hand, the assumptions that were made regarding occupancy, use, and environmental standards when the building was originally designed usually incorporated large safety factors. A careful survey of a building that has been in operation for a while permits a much more precise definition of the activities which take place there, such as the numbers of people actually involved in each space and environmental levels which their activities actually require. It also permits an accurate determination of where energy is actually consumed within the building. This information will make obvious the areas which are readily susceptible to significant energy savings.

A reinvestigation of the design criteria for the mechanical systems of buildings is essential. These include areas served, amount of time service is provided, and the levels at which the service is provided. The adjustment of environmental systems to reflect these observed conditions will usually produce greater savings than efforts to optimize the outputs of these systems. This is particularly true in the case of newer buildings which may have relatively efficient equipment, but were designed to extravagant standards. In many cases, the modifications to reflect actual needs require little or no capital expenditure and can be carried out by simply adjusting the existing control mechanisms or deactivating unnecessary components.

No two buildings are ever alike, for even if stamped from the identical mold, their location and the manner in which they are operated are bound to be different. Nonetheless, all buildings have three fundamental components which affect energy use: energized systems, nonenergized systems, and human systems.

Energized systems are those which consume energy directly. Typical energized systems include those used to produce heating, ventilating, cooling, humidification, dehumidification, lighting, hot-water heating, interior conveyance, food cooking, waste handling, as well as various pieces of equipment such as typewriters, computers, and television.

Nonenergized systems are those which do not consume energy directly, but which do affect the amount of energy that the energized systems must expend to get their job done. Typical nonenergized systems include walls, windows (glazing), floors, roof, ceiling, doors, etc., as well as weather, landscaping, siting, and similar factors.

Human systems comprise those persons who somehow have an impact on when and in what quantity energy is consumed. These persons include administrative personnel, operating and maintenance persons, as well as tenants and visitors.

In order to effectively proceed with an energy conservation program for a given building, it is necessary to assemble as much data as possible about the actual condition and circumstances of each system. Sources of these data include:

1 Documents on file at the building
2 Records of the design architects and engineers
3 Records of outside maintenance firms
4 Utility and fuel suppliers

The data are then processed to provide:

1 A picture of the physical structure of the building, including its equipment and the way in which it is operated
2 A statement of energy consumption, including quantity and type of fuels consumed and the purposes for which they were expended

THE PHYSICAL SURVEY

The first step is to assemble the basic facts of the physical plant. It is especially important to obtain the cooperation of maintenance and operating personnel early in the project. They, in large measure, will be implementing many of the changes which will be required. In most cases, these changes will mean altering what have become habits over many years. Some personnel will regard the investigation as a questioning of their ability, while others will welcome the consultant as an ally in convincing management to implement changes sought by operating personnel for a long time.

Whoever conducts the survey first must obtain a copy of the architectural, mechanical, and electrical design drawings and specifications to become familiar with the building's configuration and design, electric and mechanical systems, and equipment layout, operation, and control. If such drawings are not available, it may be necessary to develop single-line diagrams of existing mechanical and electric systems. The surveyor also should be given access to any written maintenance and operating procedures manuals supplied by equipment manufacturers or original building design professionals. The surveyor also should be advised of any planned building modernization programs and their applications. Much of this information can be recorded in a form similar to Fig. 8-1*a*, *b*, and *c*. Supplementary notes, usually spoken into a tape recorder, are taken as the survey team moves about.

Those architects who are involved in alteration or modernization work are painfully aware that record documents are not always accurate; and every effort must be made to update the data so that they reflect reality.

ENERGY CONSUMPTION

In order to develop data on past and present patterns of energy consumption, it is necessary to get copies of the fuel bills. The records required are usually on file at the institution or at the utility or fuel company from which energy is purchased. Copies of the actual bills themselves are needed; a bookkeeper's transcription is of no value because it rarely includes all the background data needed for analysis.

At the very least, energy consumption data are required for the year prior to that in which the energy study is being conducted. Ideally, if more years are available (up to five), the data will be more meaningful with respect to changing performance of the structure under different weather conditions.

Information taken from the bills is recorded on a form similar to Fig. 8-2. The form records data on a 12-month basis corresponding to a regular calendar year. Energy use should be taken from bills representing the greatest number of days in the month in question.

Electricity (kWh): Enter electric energy consumed (kWh) as given on the electric bill.

Electricity demand (kW): Enter both billed and actual demand given on the electric bill.

Electricity total cost (dollars): Enter total net cost (no late payment penalty) given on the electric bill.

Surveyed by: ____________

Survey Date: ____________

I. GENERAL INFORMATION

IDENTITY:

Name of building ____________

Address ____________

Type(s) of occupancy ____________

Name of owner(s) ____________

Person(s) in charge of building ____________

PHYSICAL DATA:

No. of floors ________ Gross Conditioned floor area ____________

Conditioned floor area, special use:

________ Ft² Computer Facility ________ Ft² Shops

________ Ft² Cafeteria ________ Ft² Other (describe ________

________ Ft² Bank ____________

Construction type:

Walls (masonry, curtain, frame, etc.)

N ____ S ____ E ____ W ____

Roof:

Type: Flat ________ Color: Light ________

Pitched ________ Dark ________

Windows:

Type: ________ Fixed Sash ________ double hung: ________ Casement ________

Doors:

Door types and numbers North ________ No ________ Type ________

East ________ No ________ Type ________

West ________ No ________ Type ________

South ________ No ________ Type ________

Type designation: ________ 1-single; 2-vestibule; 3-revolving

Glazing:

Exposure	*Type	%Glass/Exterior wall area
N		
S		
E		
W		

*Type: Single, double, insulating, reflective, etc.

FIG. 8-1a Architectural survey form. *(NECA, "Total Energy Management.")*

Electricity cost per unit (cents/kWh): Divide total net cost by total consumption (kWh). This value provides a quick check to determine billing errors and also provides cost escalation trends for electric energy. Record as cents per kilowatt hour.

Purchased Steam (Mlb): Enter steam energy consumed in thousand pounds as given on the steam bill.

Steam demand (lb/H): Enter actual and billed demand given on the bill.

Glass shading employed outside (check one):

Fins ____________ Overhead ____________ None ______________ Other ______________

Glass shading employed inside (check one):

Shades ________ Blinds ________ Drapes, open mesh __________

Drapes opaque ________ None ________ Other ________

Solid Exterior Surfaces:

Description of Surface	Thickness of Insulation (inches)	Type Insulation	Appropriate Surface Area (Ft^2)
Ceiling(s)/roof(s)			
Floor(s)			
Wall(s)			

Air Leakage (check as appropriate):

Structural damage (through the wall/ceiling cracks) ______________________________

Broken or defective windows ______________________________

Ventilation/Exhaust openings that remain open all the time ______________________________

____

Tightness around windows and doors ______________________________

____

Note other major observed air leaks ______________________________

____

____

FIG. 8-1*b* Architectural survey form. *(NECA, "Total Energy Management.")*

The difference between the billed and the actual demand is similar to the difference involved for electric energy. Usually the cost of steam is controlled by consumption, not demand. By far the best way to reduce steam costs is to cut steam consumption.

Steam total cost (dollars): Enter total net cost given on the bill.

Steam cost per unit (dollars/Mlb): Divide the total net cost by total consumption. Record as dollars per 1000 lb. This value provides a quick check to determine billing errors and also provides cost escalation trends for steam energy.

II. BUILDING OCCUPANCY AND USE

	Weekdays		Saturdays		Sundays, Holidays	
	No. Of Occupants	Period of Occupancy	No. Of Occupants	Period of Occupancy	No. Of Occupants	Period of Occupancy
General Office						
Custodial operation						
Maintenance operation						
Computer operation						
Cafeteria						
Banks						
Retail Store						
Other (describe)						
______						
______						
______						

III MAINTAINED INDOOR ENVIRONMENTAL CONDITIONS

	Occupied hours				Unoccupied hours			
	Temp (°F)		Rel. Humidity (%)		Temp (°F)		Rel. Humidity (%)	
	Summer	Winter	Summer	Winter	Summer	Winter	Summer	Winter
General office areas								
Computer Rooms								
Special process areas								

IV SYSTEMS AND EQUIPMENT DATA

HEATING, VENTILATING AND AIR CONDITIONING SYSTEMS:

(BY MECHANICAL CONSULTANTS)

FIG. 8-1c Architectural survey form. *(NECA, "Total Energy Management.")*

Fuel Oil (Gal): Enter amount of oil actually consumed during the month. Normally this amount must be determined by measurements recorded at the building during the month and is not available on bills (which correspond to deliveries).

Oil total cost (dollars): Multiply per-gallon cost on the latest bill by the amount of oil used during the month.

Oil cost per unit (dollars/gal): Enter per-gallon cost on the latest bill.

ENERGY MANAGEMENT FORM

Building ______________________________

Month	Heating Deg. Days	Cooling Deg. Days	Electricity						Purchased Steam						Fuel							Total Energy Cost
			KWH	KWH/ Deg. Days	KW Demand		Cost		M (lbs.)	M (lbs.)/ Deg. Days	Lbs./hr	Demand	Cost		Oil			Gas ☐ Coal ☐ Other ☐			Fuel/ Deg. Days	
					Actual	Billed	Total	Per Unit			Actual	Billed	Total	Per Unit	Quant. (Gal.)	Cost		Quant.	Cost			
																Total	Per Unit		Total	Per Unit		
1	2	3	4	5	6	7	8	9	10	11	12	13	14	15	16	17	18	19	20	21	22	23
Jan.																						
Feb.																						
March																						
1st Quarter																						
April																						
May																						
June																						
2nd Quarter																						
July																						
Aug.																						
Sept.																						
3rd Quarter																						
Oct.																						
Nov.																						
Dec.																						
4th Quarter																						
Total Per Year																						

FIG. 8-2 Annual energy consumption. *(New York State Energy Office.)*

Gas energy (Mcf): Enter gas consumed in thousand cubic feet (Mcf) given on the bill.

Gas total cost (dollars): Enter total net cost given on bill.

Gas cost per unit (dollars/Mcf): Divide total net cost by total gas consumed. Record as dollars per thousand cubic feet.

Other fuel(s): Enter appropriate values.

Total energy cost per month (dollars): Add across all energy types for each month. Enter total for each month.

Total energy per year per energy type: Add amount used each month in each energy type. Record in appropriate commercial units.

Total cost per year per energy type (dollars): Add monthly costs for each energy. Record for each energy.

Total energy cost per year (dollars): Add monthly energy costs. Enter for the year.

Note: While tabulating energy consumption, it is often desirable to verify the basis on which utilities are basing their bills. All utilities have different rate schedules which apply to different consumer situations, and they are under no obligation to select the most economical schedule from the consumer's point of view. (Consultants specializing in billing verification are available to assist the energy consultant in this work.)

The collection of energy consumption data is important for several reasons:

1 It provides the quantitative basis for the overall energy management program.

2 It provides clues to energy waste.

3 It indicates the greatest energy-saving potentials.

4 It provides year-to-year comparisons to keep track of energy saved.

5 It monitors overcharges in energy bills.

Consider for a moment a hospital that has no central air conditioning and uses oil for heating.* In this hospital the major uses of oil could be heating, kitchen, laundry, and sterilization. All these uses are essentially independent of the weather except for heating. Oil consumption can be expected to increase during winter ("variable load") and reduce to some minimum during summer ("base load").

Oil use for a typical hospital is shown in Fig. 8-3. Note that even during summer months, approximately 13,000 gal of oil was consumed (the base load). During the month of January, 35,000 gal was used. Thus, the oil use identifiable for heating is

$$35{,}000 - 13{,}000 = 22{,}000 \text{ gal} \qquad \text{variable load}$$

*This section is adapted from "Practical Energy Management in Health Care Institutions" prepared for the FEA by the University City Science Center, Philadelphia, Pennsylvania. Copywright © 1977 by Blue Cross of Greater Philadelphia.

The variable load is the apparent energy for heating needs. Dividing this value by the degree-days in the month results in a useful heating efficiency factor for the entire building.

This factor (gallons of oil per degree-day per month) can vary greatly from month to month, as shown in Fig. 8-4. It is usually lowest (best) in the middle of the heating season. At this time the heating plant is at its maximum operating efficiency. The curve in Fig. 8-4 is always U-shaped. Nevertheless, the unusually high values of the efficiency factor in Fig. 8-4 during intermediate months (spring and fall) suggest that improvement of heating controls could save energy.

When cooling for air conditioning is provided primarily by electrically driven chillers, summer electrical consumption minus the base load reflects the electricity used for cooling. A building of this type is shown in Fig. 8-5. This building

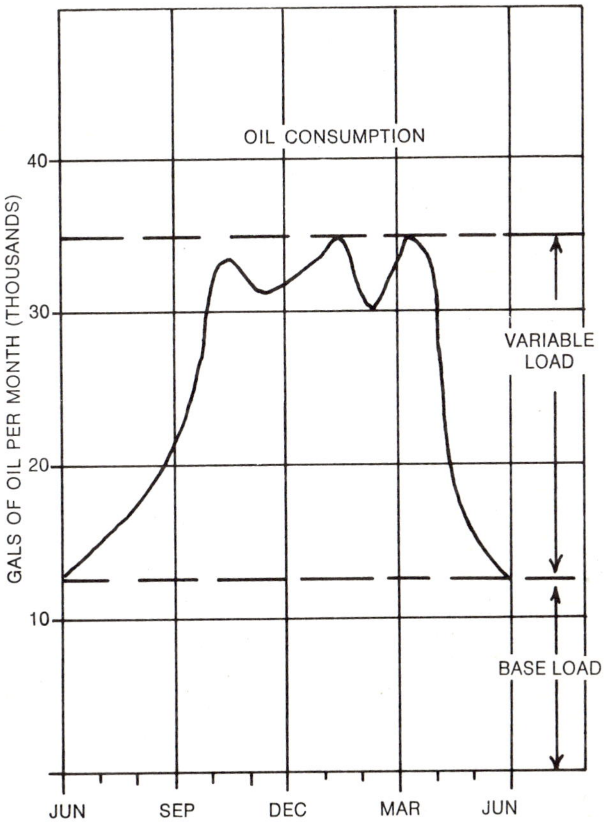

FIG. 8-3 Annual oil consumption (Mgal/mo). *("Practical Energy Management in Health Care Institutions." Copyright 1977 by Blue Cross of Greater Philadelphia.)*

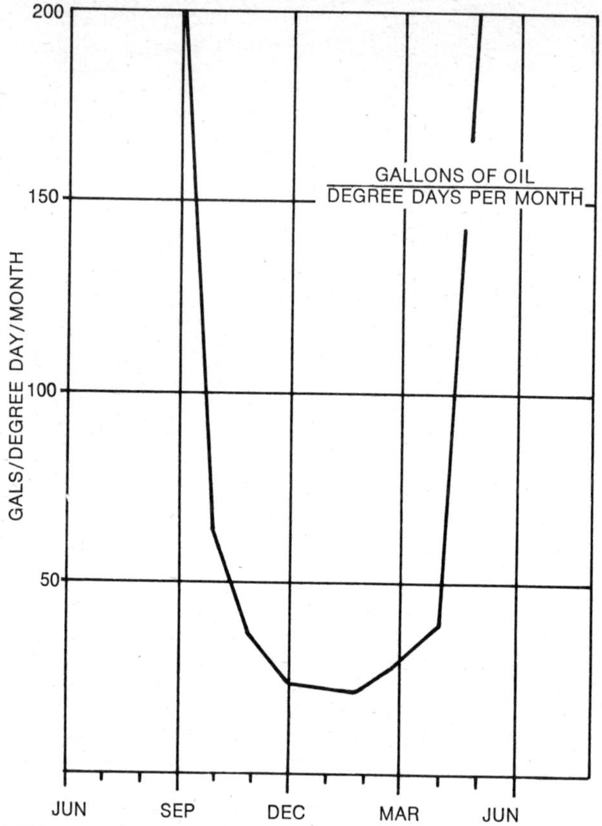

FIG. 8-4 Annual oil consumption (gal/degree-day · mo). *("Practical Energy Management in Health Care Institutions." Copyright 1977 by Blue Cross of Greater Philadelphia.)*

also buys steam. Note how the steam and electricity peak at different times of the year. Clearly, steam is the "heating energy" and electricity is the "cooling energy."

It is important to recognize that in buildings with air conditioning the base load reflects not only weather-independent uses of energy, but also energy to operate auxiliaries (such as fans and pumps in environmental control systems) and wasted energy (such as year-round simultaneous heating and cooling of air).

Identifying the base load is somewhat more difficult in buildings that have both electric heating and/or steam absorption chillers for air conditioning. The practical approach is to define the base load as the lowest monthly consumption of energy even though it may occur during spring or fall. Once the base load is identified, each month's variable load can be easily calculated.

Figures 8-6 and 8-7 show the same hospital pictured in Fig. 8-5. Figure 8-6 superimposes heating degree-days per month on the consumption data given

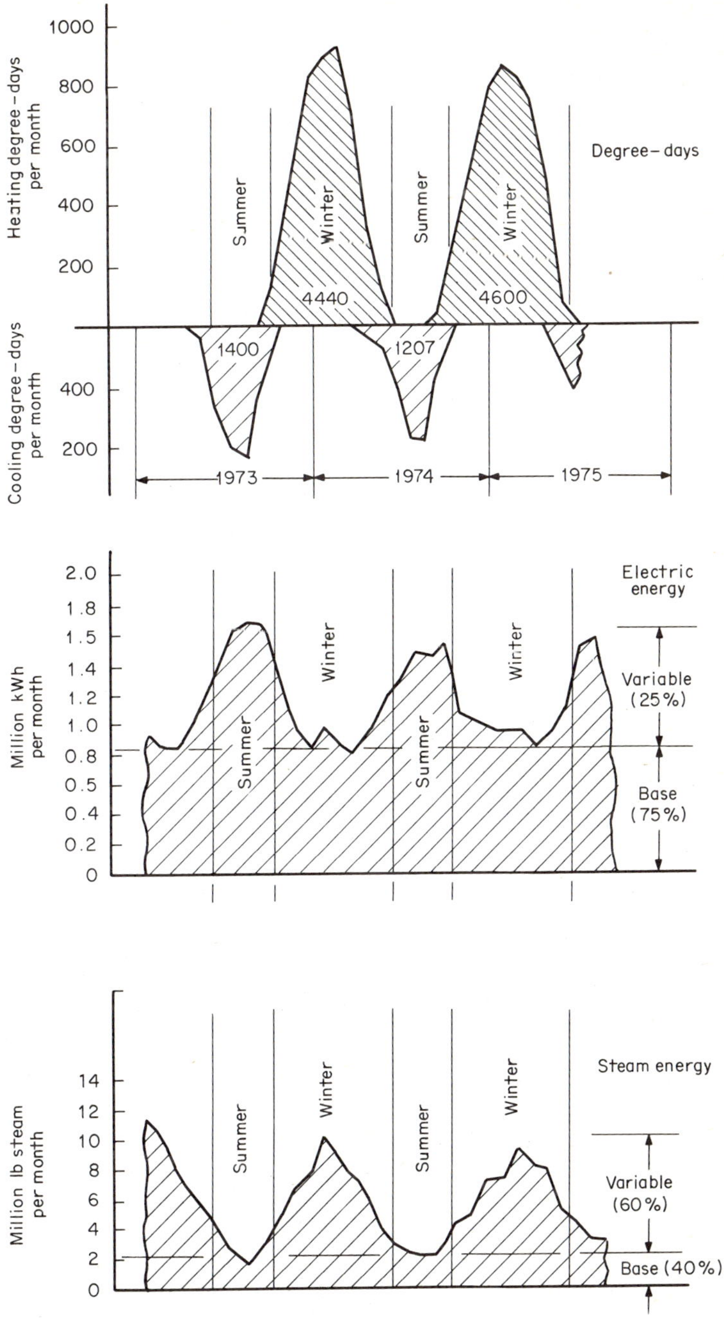

FIG. 8-5 Yearly variation in energy consumption per month. *("Practical Energy Management in Health Care Institutions." Copyright 1977 by Blue Cross of Greater Philadelphia.)*

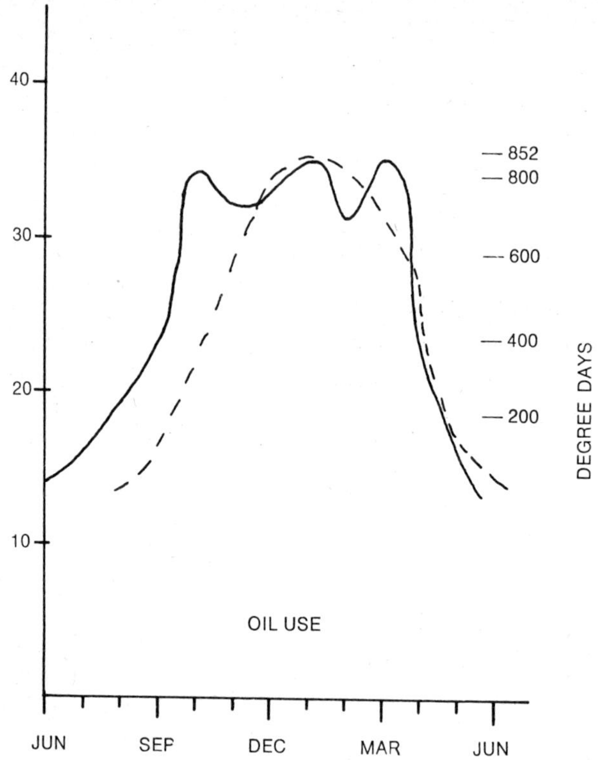

FIG. 8-6 Yearly oil consumption. *("Practical Energy Management in Health Care Institutions." Copyright 1977 by Blue Cross of Greater Philadelphia.)*

in Fig. 8-3. The degree-days are plotted as a dashed line. Their scale is adjusted so that zero falls on the consumption base load and the peak consumption and peak degree-days coincide. That is, the heating degree-days are normalized to the consumption curve. Whenever the consumption curve is higher than the degree-day curve, energy is wasted. Note that the space between the two curves in Fig. 8-6 indicates the same results suggested in Fig. 8-3. Figure 8-7 shows electricity. In this case, the dashed curve represents cooling degree-days per month. This building is using relatively little air conditioning. Therefore, it should have, and does have, a relatively high electric base load. Or, to put it another way, there is no large summer peak for air conditioning.

Figures 8-8 and 8-9 show energy consumption for a building with primarily through-the-wall air conditioners. Some air conditioners have hot-water heat, and a few have electric heat. The peak in oil consumption for August (Fig. 8-8) was believed to have been caused by defective temperature controls. A similar peak did not occur the following year after the controls were repaired.

Figures 8-10 and 8-11 show energy consumption for a building with several centrifugal (electric) chillers for air conditioning. A change in system operation to shut down chillers during intermediate months appears to be an energy-saving possibility.

Figures 8-12 and 8-13 represent a complex composed of a few large buildings, some with systems using 100 percent outside air.

Figures 8-14 and 8-15 show a building with a relatively well-controlled but high heating-energy peak (high variable load, low base load). This condition suggests that the building has systems using large quantities of outside air (compare with Figs. 8-12 and 8-13).

Clearly, simple plots of energy consumption contain a great amount of useful information. They provide a powerful, overall analysis of energy use.

If possible, plot at least three years' energy consumption. Use simple plots (five squares to the inch, one month per square) of monthly consumption for each major energy. Also plot the cooling and heating degree-days each month. Align the plots as shown in Fig. 8-5 and determine energy used for cooling (base load and variable load) as well as energy used for heating (base load and variable load).

Examine the individual plots of each major energy. Look first for any anomaly—the "zig" in the curve. Review again the oil consumption of Fig. 8-8.

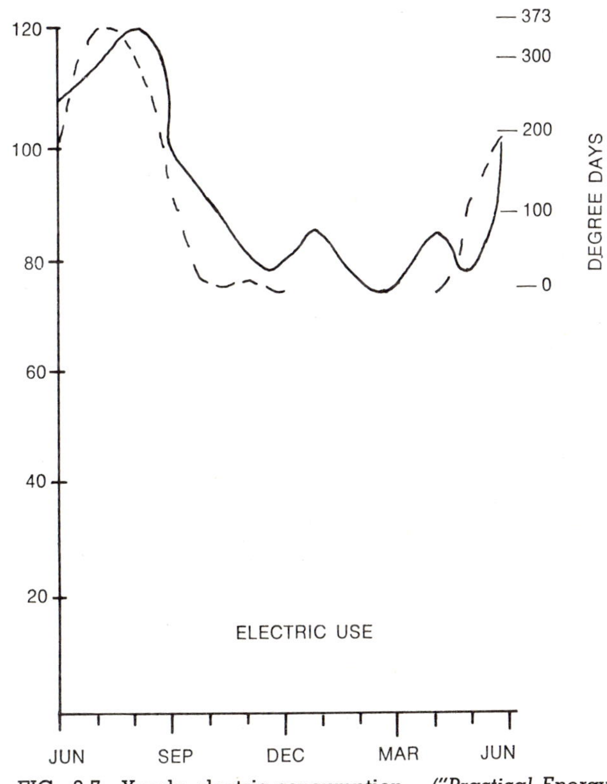

FIG. 8-7 Yearly electric consumption. *("Practical Energy Management in Health Care Institutions." Copyright 1977 by Blue Cross of Greater Philadelphia.)*

Weather data did not show a cool August, but there is a high heating peak in August. The exact cause is not known. The building initiated a maintenance program to upgrade preheaters and repair controls at the end of the year. The following year's consumption did not show a similar peak.

Examine the overall profile of each major energy. Peaks and valleys should occur during the same month each year. Identify exceptions to the profile. Perhaps a portion of the building was shut down. Perhaps the weather was different. Find out. You may be seeing the result of a defective system.

Determine why there are differences in the magnitude of peaks and valleys. Relate these to the operation of heating and cooling systems as well as to the overall building operation. A series of increasing peaks may indicate additions to the building or progressive deterioration of system performance.

Plot normalized degree-day curves superimposed on energy consumption curves (as shown in Figs. 8-6 through 8-15). Examine the difference between the two curves. The difference represents wasted energy.

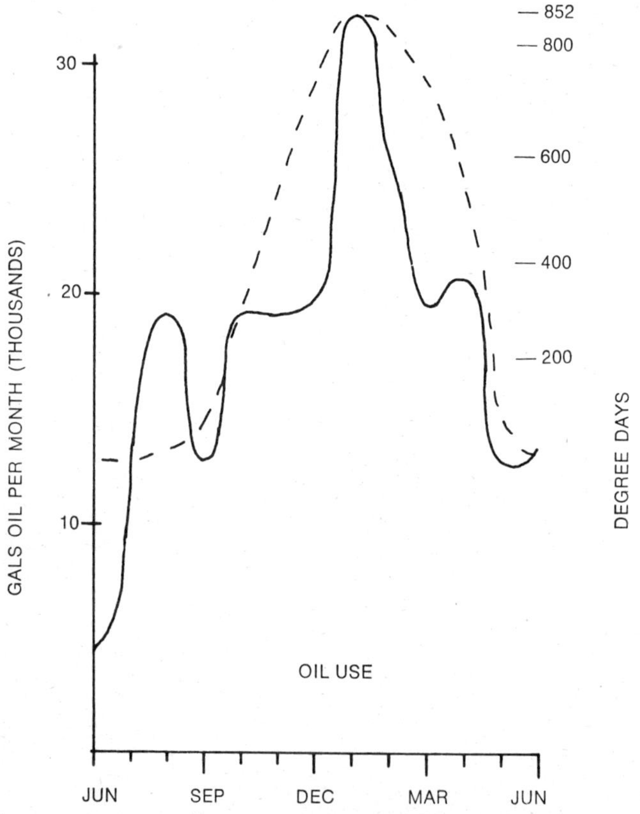

FIG. 8-8 Yearly oil consumption for a single building with unit air conditioners. *("Practical Energy Management in Health Care Institutions." Copyright 1977 by Blue Cross of Greater Philadelphia.)*

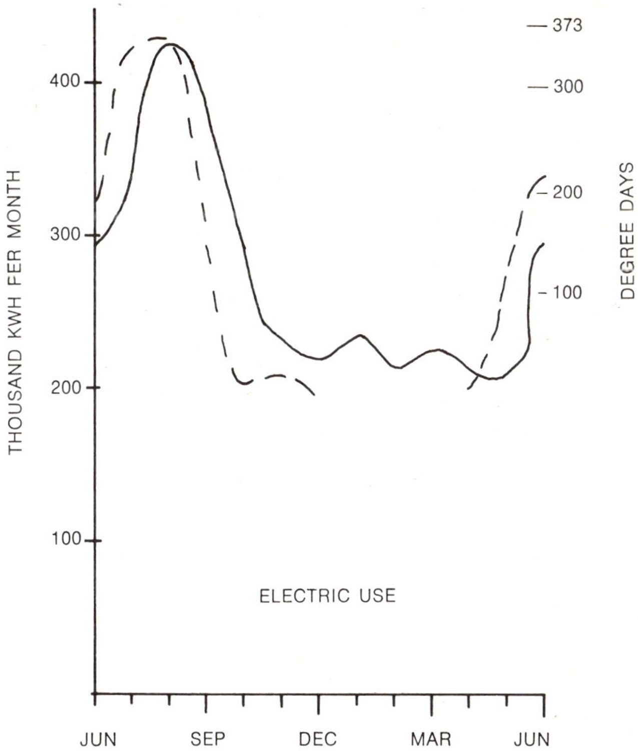

FIG. 8-9 **Yearly electric consumption for a single building with unit air conditioners.** *("Practical Energy Management in Health Care Institutions." Copyright 1977 by Blue Cross of Greater Philadelphia.)*

Plot heating variable load per degree-day per month. Also plot cooling variable load per degree-day per month. Widely varying values indicate an opportunity to save energy by improving controls.

THERMOGRAPHY

Every object whose surface temperature is above absolute zero (−273°C) radiates electromagnetic energy which is a function of its surface temperature. The intensity of this radiation varies along the wavelengths of the electromagnetic spectrum from the visual to the infrared ranges.

By using highly sensitive infrared camera systems, it is possible to observe and record the thermal picture of the temperature differences on a surface such as the exterior of a building. The resulting heat picture, known as a thermogram, shows these temperature differences in either a color picture or black and

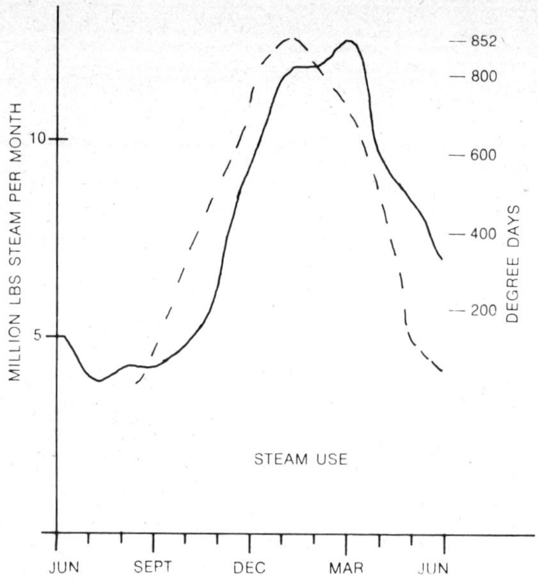

FIG. 8-10 Yearly steam consumption for a single building with several chillers. *("Practical Energy Management in Health Care Institutions." Copyright 1977 by Blue Cross of Greater Philadelphia.)*

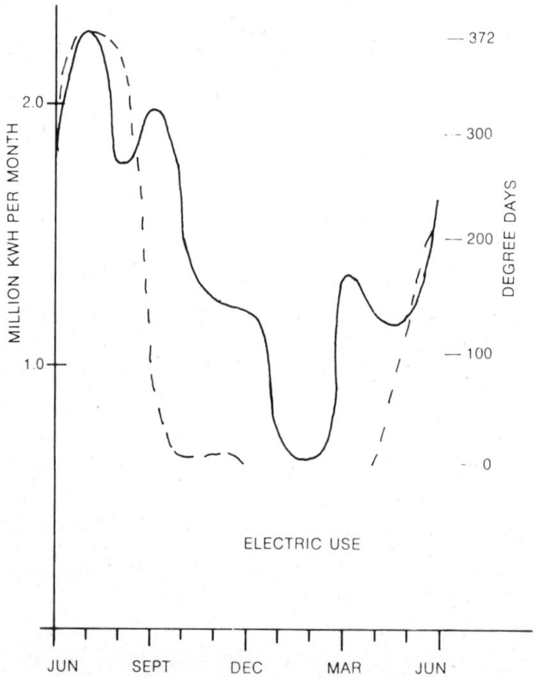

FIG. 8-11 Yearly electric consumption for a single building with several chillers. *("Practical Energy Management in Health Care Institutions." Copyright 1977 by Blue Cross of Greater Philadelphia.)*

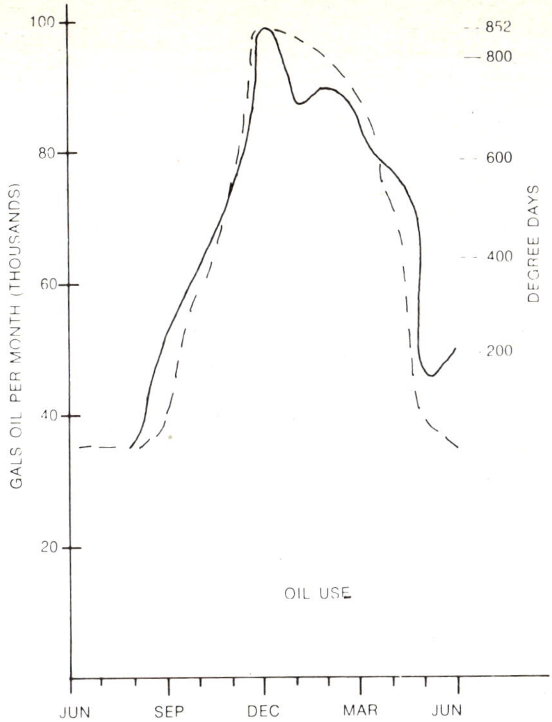

FIG. 8-12 Yearly oil consumption for a multibuilding complex. *("Practical Energy Management in Health Care Institutions." Copyright 1977 by Blue Cross of Greater Philadelphia.)*

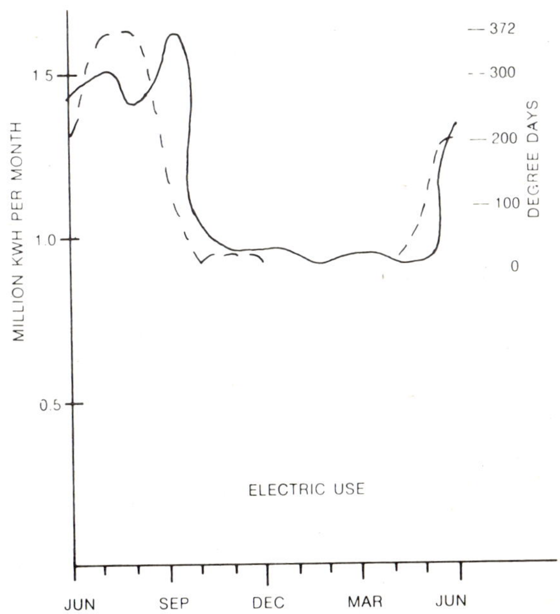

FIG. 8-13 Yearly electric consumption for a multibuilding complex. *("Practical Energy Management in Health Care Institutions." Copyright 1977 by Blue Cross of Greater Philadelphia.)*

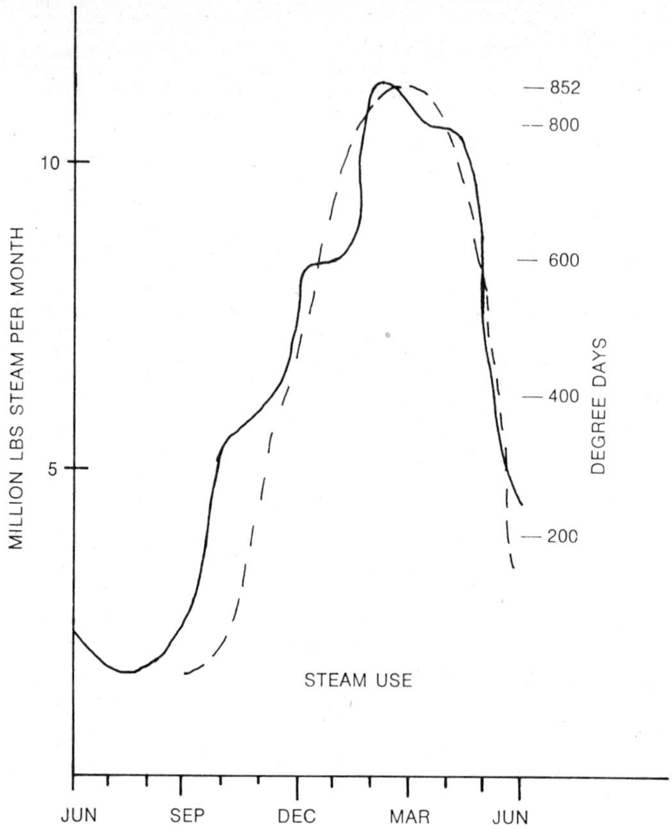

FIG. 8-14 **Yearly steam consumption for a single building with a high heating peak.** *("Practical Energy Management in Health Care Institutions." Copyright 1977 by Blue Cross of Greater Philadelphia.)*

white (both can be recorded on Polaroid attachments). In color, hotter surfaces appear at the red end, cooler at the blue. In black and white, the hotter the surface, the lighter the image.

Figure 8-16, taken with a conventional camera, shows a picture of a school office window with an air conditioner in the upper sash. Figure 8-17 shows a thermograph of the same window; it reveals two replacement panes of glass leaking heat and inadequate insulation around the air conditioner.

Since the surface of a building tends to acquire the ambient temperature of its surroundings, the reason different parts of the exterior exhibit temperatures is primarily due to heat loss from the interior in winter (and the reverse in summer). In effect, the thermogram provides a visual picture of how and where heat and cold are passing through the structure.

Thermography is valuable for two reasons. It shows where heat is really being lost (as opposed to theoretical design assumptions); and it shows this loss dramatically and therefore acts as a powerful incentive to push management into implementing energy conservation work. The equipment is not cheap

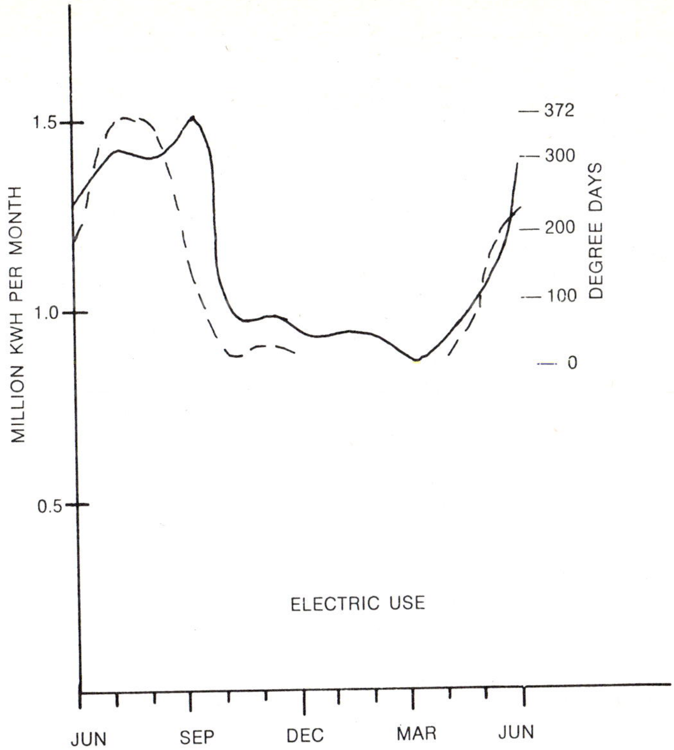

FIG. 8-15 Yearly electric consumption for a single building with a high heating peak. *("Practical Energy Management in Health Care Institutions." Copyright 1977 by Blue Cross of Greater Philadelphia.)*

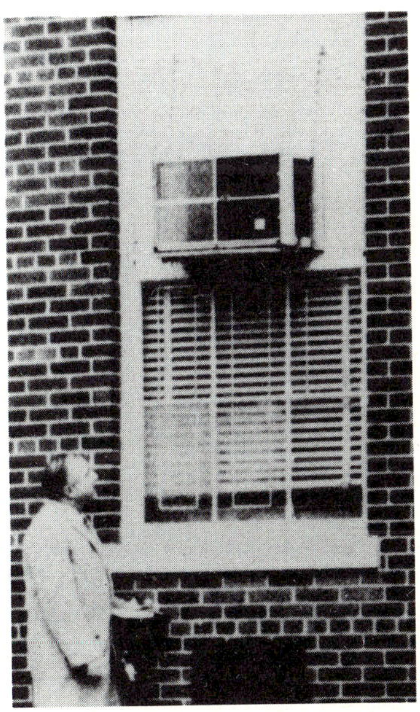

FIG. 8-16 Conventional photograph.

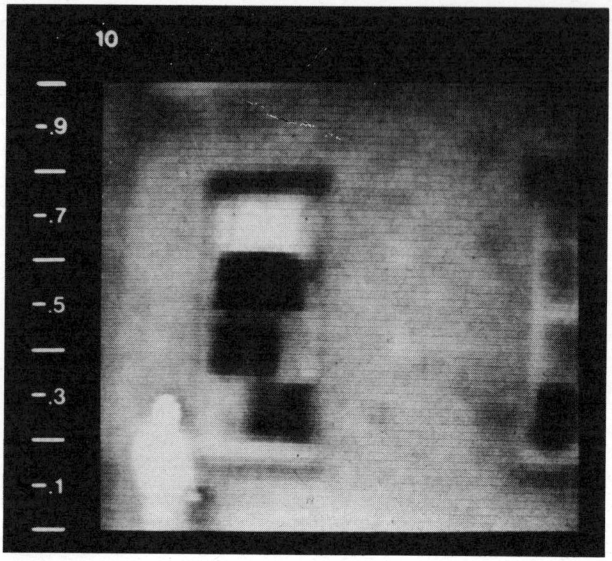

FIG. 8-17 Infrared thermograph.

(upward of $50,000), but both equipment and operators can be rented on a daily basis. One day's or two days' work is usually sufficient for most buildings to record the data and supply the photographs (thermograms). Thermograms can also provide interesting insights into the condition of operating equipment (where overheating reflects deterioration), as well as being a construction supervision tool through their ability to provide clues to construction ordinarily hidden within walls, etc., which transmit heat differently and thus provide a picture akin to an x-ray.

EQUIPMENT-INTENSIVE STRUCTURES

There are buildings where owner-operated equipment may far outweigh the energy consumption of the building itself. An obvious example of this might be an industrial plant whose manufacturing machines consume large amounts of energy. Conservation measures for this equipment and its interaction with the building structure usually require specialized consulting backup and are not within the scope of this book.

Assistance should again be sought when the energy analyst is dealing with groups of buildings such as a college campus, rather than with individual structures alone. Distribution losses and the interaction of one structure with another are but two of many considerations in dealing with building complexes. The basic analysis covered in this chapter will still be followed, but the ability to distinguish each part of the whole usually requires special metering, monitoring, and interpretation.

Chapter 9

Retrofitting Existing Buildings: Exploration and Analysis

Having gathered the data that describe the energy anatomy of the structure, we are ready to begin an analysis leading to improvement. The most successful energy conservation programs generally move from the most elementary steps to levels of increasing sophistication. Successful programs are based on the initiation of the most obvious conservation steps those that may require little capital investment but some effort in retraining people and reshaping attitudes.

In order to understand the phases, or the multistep approach, to effective energy management, it is useful to separate energy modifications into three categories, based on consumption reduction and payback: quick fix, refit, and systems convert. As noted in Table 9-1, each category of modification, when added to the work preceding it, offers a greater opportunity for total consumption reduction, and each involves an increasing rate of captial expenditure.

Quick-fix modifications can be defined as measures of energy consumption reduction that achieve rapid and significant energy savings at negligible cost or at a cost that can return an investment within a year. These modifications can normally be accomplished by operating and maintenance personnel. Conservation measures such as the adjustment of temperatures and lighting levels will quickly eliminate obvious waste, resulting in energy reductions of at least 10 percent for most institutions. Minimal preparation is usually required for implementing these quick-fix measures.

Refit modifications are those measures that require modest capital investment, can usually pick up an

Table 9-1 Savings and payback

	Energy Savings (%)	Payback Period
Quick fix	10–20	Up to 1 year
Refit	10–15	1 year to 3 years
Systems convert	10–10+	2 years to 5 years
Accumulated	30–45+	

additional 10 to 15 percent in energy savings, and have paybacks in one to three years. These include such measures as reduction of transmission heat losses and gains and modification of space characteristics.

Systems-convert modifications are generally sophisticated and expensive. They generally yield an additional 10 percent savings in energy and involve paybacks of two to five years. Normally, these modifications are implemented after thorough quick-fix and refit programs have been initiated.

In looking over existing buildings, the analyst will often get complaints about comfort conditions bothering occupants of the structure. Tying in correction of these conditions as part of the conservation program often proves a way to spur implementation of the conservation work.

THE WALK-THROUGH SURVEY

The survey team generally includes the architect, a conservation-oriented mechanical engineer, and the building's operating engineer (preferably alone, since the presence of other administrators tends to restrict conversation). Assuming that the background data described in Chap. 8 have previously been supplied and digested, the basic tools for the walk-through include pen or pencil and paper, minitape recorder, a camera (with flash), and a light meter.

It cannot be overemphasized that every effort should be made to ensure that in-house personnel do not look upon the work of outside consultants as being a challenge to their own ability or credibility, since their cooperation is absolutely essential to the success of the program.

In addition to the physical examination that takes place during the walk-through, there is also good opportunity to chat with operating personnel and become familiar with their thoughts on improvement as well as their general understanding of the systems in their charge. Information with respect to building modifications likely in the near future, as well as any conservation work already on order, can also be obtained from them.

Every structure will have its own salient points. Usually the walk-through will deal with airtightness of the building and how infiltration can be reduced; transmission characteristics and how they can be modified; ventilation system operation and controls and how they can be improved; operating procedures and how they may be contributing to excessive energy consumption; heating and cooling equipment, maintenance, and controls and methods of improving their efficiency; lighting and lighting levels and how they can be modified; etc. The survey should result in a picture that details deficiencies of energized, nonenergized, and human systems which have led to excessive energy consumption.

As the survey team moves about the building, many ideas will come to mind. All these should be recorded even though later analysis will inevitably cause many of them to be discarded. The analysis of these "duds" has two advantages: first, it can trigger an associated, but helpful, thought in someone else; second, the same suggestion may be made at a later time by the client and it is good to have the analysis already at hand.

Many businesses and institutions belong to institutes or trade associations which have prepared checklists to be filled out by energy survey teams particularly geared to their own type of building. Wherever these are available, they should be used. In addition, the ASHRAE 100 Series for retrofitting existing structures (described in Chap. 1) contains many relevant points for conducting a survey.

BRAINSTORMING

With the walk-through completed (a process whose duration varies with the complexity of the structure), the time for consideration of remedial measures is at hand.

It will be recalled (Chap. 8) that three main energy systems apply to buildings: energized systems, nonenergized systems, and human systems. Energized systems are largely the province of the mechanical engineer, and, with few exceptions, architects are well advised to leave suggestions for this area of the work to their consultants. On the other hand, both the nonenergized and the human systems deal with subjects that are basic to architectural practice. Accordingly, when the survey team splits up after the walk-through to begin their studies and make their preliminary recommendations, the architect will devote most of his or her activity to the nonenergized and human systems.

The following checklist is intended to stimulate the architect's thinking in this phase of the work.

General

- Compare original design assumptions against building actuality.
- Reexamine the assumptions that were made regarding occupancy, use, and environmental standards when the building was originally designed.
- Convey this user information to your mechanical consultant to help in reinvestigating the design criteria for the building's mechanical systems, including areas served, amount of time service is provided, and the levels at which the service is provided.

- Reduce the building's dependence on mechanical systems. For example, consider the installation of operable windows for natural ventilation.
- Investigate whether components such as sunshades added to the facade can significantly reduce energy use.

Site

- Shade walls and paved areas adjacent to the building to reduce solar radiation striking the building.
- Plant deciduous trees for their summer sunshading and windbreak effects for buildings up to three stories.
- Plant coniferous trees for summer and winter sunshading and windbreak effects.
- Cover exterior walls with planting to reduce heat transmission and solar gain.
- Plant shrubs or lawn between the building perimeter and sidewalk to reduce heat buildup.
- Consolidate or, where feasible, relocate paved areas away from the building perimeter and replace with grass or other vegetation to reduce outdoor temperature buildup.
- Use ponds and water fountains to reduce the ambient air temperature around the building.
- Shelter service areas from sun and wind with screens and walls and/or by judicious planting of trees.
- Explore ways of enclosing service and delivery areas.

The Building

- Evaluate the tradeoffs between heat loss and heat gain on each exposure with particular regard to the wind and the sun.
- Evaluate the utilization of daylight for interior lighting versus heat loss and heat gains and glare problems peculiar to each exposure.
- Evaluate the beneficial effects of direct solar radiation for heating versus the adverse effects during the cooling season.
- Evaluate the extent to which alteration of the building envelope would maximize energy conservation benefits when all the tradeoffs are evaluated.
- In existing buildings of considerable mass (masonry walls) and mostly exterior space, consider no air conditioning; limit cooling loads by use of multipane glass, reduced glass area, sunshading, heat-removal light fixtures, remote-ballast light fixtures, reduced lighting, and additional wall and roof insulation.
- Review existing roof, wall, floor, and all other exterior insulation.

- Avoid thermal bridges through the exterior surfaces.
- Provide textured finish to external surfaces to increase the external film coefficient.
- Reduce heat transmission through the roof by adding a double roof and ventilating the space between.
- Provide solar shading for the walls and roof in the same areas where similar solar control is desirable for glazing.
- To minimize heat gain in summer from solar radiation, finish the walls and roof with a light-colored surface having high emissivity.
- To increase heat gain from solar radiation on walls and roof, use a dark-colored finish having high absorptivity.
- To reduce heat loss through glass windows, consider one or more of the following:

  Reduce the ratio of glass wall and window area to opaque wall area.
  Reduce the percentage of glazing on north wall to a minimum.
  Convert to double or triple glazing.
  Replace window frames that form a thermal bridge.
  Use operable thermal shutters.
- To reduce heat gain through glass, consider one or more of the following:

  Reduce the ratio of glass wall and window area to opaque wall area.
  Convert to double or triple glazing.
  Consider double glazing with a reflective surface.
  Reduce the percentage of exposed window glazing on the south wall.
  Use ventilation blinds or drapery as interior shading devices.
- Consider the use of solar controls, such as internal or external shading devices, to manipulate sunlight and achieve the maximum advantage and energy savings from it.

  Solar shading devices used to reduce heat gain in the summer are most effective when located on the exterior of the building and particularly effective when movable.

  Sun controls are most effective when designed for each specific orientation. Horizontal shading devices, such as awnings, prevent solar gains when installed on southern exposures. On east and west walls, a combination of horizontal and vertical devices is needed to prevent solar gains. For low-rise structures, vertical sun controls might simply be trees or shrubs.

  Venetian blinds act as adjustable solar screens and can cut solar gains considerably.

  When closed, draperies are sometimes more effective at controlling solar gains than tinted glass.

  Exterior louvered solar screens, as well as interior shutters, can prevent excessive heat loss in addition to controlling solar gains. Shutters can be equipped with devices to seal windows at a preset time or temperature.

  Color is important in all the above treatments. Where cooling is a problem, venetian blinds and the inside and outside of shutters should be light in color to reflect heat and light away from the building. Where heating and/or light restriction is desired, the use of dark draperies or dark shades makes sense.

- To minimize heat gain in summer resulting from solar radiation, consider one or more of the following:

  Finish roofs with a light-colored surface having high emissivity.

  Convert to reflective surfaces.

  Install roof sprays.

  Install roof ponds.

- To allow the use of natural light in cold zones where heat losses are high energy users, consider operable thermal barriers.

- To take advantage of natural daylight within the building and reduce electric energy consumption, consider the following:

  Locate windows high in the wall to increase reflection from the ceiling and reduce glare effect on occupants.

  Control glare with translucent drapes operated by photo cells.

  Provide exterior shades that eliminate direct sunlight but reflect light into occupied spaces.

  Slope vertical wall surfaces so that windows are self-shading but reflect light into occupied spaces.

- In climatic zones where outdoor air conditions are suitable for natural ventilation, provide operable windows. Where applicable, allow cross ventilation through occupied spaces.

- To reduce air infiltration quantities, consider one or more of the following measures:

  Be sure fixed-glass facades are adequately gasketed and/or caulked.

  If possible, use permanently sealed windows to reduce infiltration in climatic zones where this is a large energy user.

  Reduce crack areas around doors, windows, etc., to a minimum.

  Provide all external doors with weatherstripping.

  Be sure operable windows have sealing gaskets and cam latches.

  Verify that the mechanical engineer has checked exhaust air for reduction.

  Provide all entrances with unheated and uncooled vestibules or revolving doors (check code requirements).

  Provide vestibules with self-closing weatherstripped doors to isolate them from the stairwells and elevator shafts.

  Seal all vertical shafts.

  Provide windbreaks for ventilation louvers.

- Use open planning with operational and mechanical systems geared accordingly for more efficient lighting, simpler, more efficient ventilating, and more efficient heating.

- Open planning allows excess heat from interior spaces to transfer to perimeter spaces that have a heat loss.

- Use corridors as heat-transfer buffers and locate against exterior walls.

- Group service rooms as a buffer and locate at north wall to reduce heat loss, or the south wall to reduce heat gain, whichever is the greatest yearly energy user.
- Locate rooms with high process heat gain (e.g., computer rooms) against outside surfaces that have the highest exposure loss.
- Equipment rooms located on the roof allow more direct duct and pipe runs, thus reducing power requirements.
- Processes that have temperature and humidity requirements different from normal physiological needs should be grouped and served by one common system.
- Rooms can be grouped so that the same ventilating air can be used more than once, by operating in cascade through spaces in decreasing order of priority, e.g., office, corridor, toilet.
- Spaces of a related function located adjacent to each other on the same floor reduce the use of elevators.
- Offices frequented by the general public located on the ground floor reduce the use of elevators.
- Lower ceiling heights reduce the exposed wall surface area and the enclosed volume. They also increase illumination effectiveness.
- Deep ceiling voids allow the use of larger duct sizes with low pressure drop which will reduce HVAC requirements.
- Judicious use of reflective surfaces such as white ceilings can enhance the effect of natural lighting and increase the yearly energy saved.
- Use carpeting for thermal comfort, particularly over slabs on grade.
- Do not heat parking garages.

Not every suggestion listed applies to every situation that the analyst will encounter. Indeed, some of the suggestions contradict others inasmuch as they are intended to cover different circumstances. But as the architect mulls over certain ideas, looks back over survey data, studies the thermographs, and reviews field notes, the architect will begin to zero in on certain possibilities.

The mechanical engineer will be going through a similar process and developing a list of possibles.

At this point both the architect and the engineer should meet to discuss their tentative recommendations and have this discussion before either has gone through any exhaustive analysis of any idea. The reason for the meeting is to eliminate those possibilities which either consultant knows from experience cannot be made to work from an architectural, an engineering, or a financial point of view.

Looking at the architect's suggestions, the engineer will probably be able to say that some ideas will never save enough energy to pay off their implementation cost; and the architect, looking over the engineer's list, will veto many suggestions felt to cause more architectural damage than their energy savings warrant.

WINNOWING

Of the ideas that survive this first cut, a more exhaustive analysis is undertaken by each office. In general, this phase consists of taking each idea and describing it in detail, estimating its cost, computing its energy savings, and calculating its payoff time in the simplest possible way by dividing its cost by its annual fuel saving.

For example, suppose that a walk-through of a hospital has shown that the main lobby and its adjacent main cafeteria are little used at night. This might have been listed on the initial collection of possibles as "lobby-cafeteria shutdown." Once it has been decided that additional study is worthwhile, the description might be expanded to the following:

Shutdown of AC-1 Cateteria, Main Lobby Unit (West Wing)

This AC unit serves areas that could be shut down at night. Basically the main lobby is closed from 8:30 p.m. until 7 a.m. except for occasional use by relatives of emergency medical patients. Similarly the cafeteria, which is on the same system, is shut down for the same hours. If these areas are occasionally used by a few personnel, we don't anticipate a problem with temperature or ventilation. If it is a problem, the time clock can be reset for shorter off hours. The systems should have a night stat control to occasionally cycle fan to maintain temperature. Construction cost should include night stat, time clock, wiring, etc. This can be tried manually on a trial basis before installing controls.

Costs and probable energy savings are then computed as shown in Fig. 9-1.

Finally, the payback is calculated:

$$\frac{\$1500}{\$672 + \$1416} = 0.71 \text{ yr}$$

The same procedure is run through for each item. These data then form the basis for final recommendations. Before the final selection is made, it is a good idea to get back in touch with the third member of the survey team, the building's operating engineer. Not only is it good public and personnel relations, but the operating engineer may often spot obvious bloopers missed by the energy consultants in their survey of the structure. It also makes it possible to get some idea of which items might be undertaken out of the normal operating budget and those which will need special funding. These thoughts on where this funding might come from, or at least who are the best people with whom to discuss the subject, can be very helpful.

The final choice of recommendations is thus made by three parties: the architect, the mechanical engineer, and the building's operating chief.

In general, most private businesses will be primarily interested in those items with very short payback periods, rarely more than three years. Institutional, government, and the large corporate clients also will be interested in periods as

GOLDMAN SOKOLOW COPELAND CONSULTING ENGINEERS 212 - LE 2-9512

Page 49

C-1 - Equipment Shutdown

Electrical Savings

Using actual brake horsepower from drawings S-1 supply fan (11.72) + R-1 return fan (4.44) = 16.16 BHP, therefore: 16.16 BHP x 1.1 (10% drive loss) x .747 KW/BHP = 13.27 KW.

Hours of non-use for areas served by this unit: lobby open from 7 am to 8:30 pm approx. 14 hours, cafeteria open similar hours. Assume systems cycle approximately 10% of time to maintain space temperature.

Therefore:

Savings = 10 hrs/day x .9 (off time) x 365 days/yr. x 13.27 KW x .0325 \$/kw. hr.
= 1416 \$/year.

Oil Savings

CFM = 10,520 CFM $\quad Q_{OA} = 10,520 \times 1.08 \times 70 = 795,312$

Using Formula From ASHRAE Guide $\quad X = \dfrac{795,312(27)}{65} = 330,360$

$D = 5219 \quad \Delta t = 70 - 0 = 70$
$h = .75 \quad v = 150,000$
$C_D = .71 \quad Cf = 1.56$

$$E = \left(\frac{Q_{OA} \times D \times 12}{\Delta t \times h \times v}\right) = C_D C_F$$

$$E = \left[\frac{795,312 \times 5219 \times 12}{70 \times .75 \times 150,000}\right] .71 \times 1.56 = 7005 \text{ GALS.}$$

Assuming Heat Wheel Efficiency of 70%, Multiply x 0.3 Thus:

7005 Gals. x 0.3 x .32 \$/gal. = \$672

Construction Cost

Timeclock, One Nite Stat and Wiring $1,500

Appendix - 3

PROJECT Long Beach Hospital Date ______ No. ______

FIG. 9-1 Probable energy savings and costs. *(Long Beach Memorial Hospital Conservation Study.)*

long as ten years since they can justify the expenditure on an investment basis rather than thinking only of annual cash flow, which is the worry of most private businesses.

Once the recommendations have been chosen, the report is prepared as described in Chap. 10.

Chapter 10

Retrofitting Existing Buildings: Final Report

This chapter presents an example of a retrofit analysis prepared by the author's architectural office. Other architects will undoubtedly alter the format to correspond with their own office procedures and will vary the presentation to conform to their findings with respect to a particular building. This example is offered therefore only to serve as a rough guide and jumping-off point for any particular project.

BACKGROUND

Long Beach Hospital is located on the south shore of Long Island in New York State. The original structure is about 50 years old and has been added to and altered many times during this period.

The author's firm has been involved in many of these alterations as well as with newer (and separate) structures for the hospital. After the OPEC oil crisis of 1973–1974 the question of reducing energy consumption began to come up more and more at meetings of the hospital's building committee, eventually leading to a commission to study what could be done with the structure. Consulting mechanical engineers were Goldman, Sokolow, and Copeland of New York City.

The study and report took about four months to accomplish, and technical time came to about 300 h.

FORMAT

The final report was 90 pages long, divided as follows:

	Page Numbers
Summary of recommendations	1–3
Description of recommendations	4–13
Thermography	14–39
Architectural observations	40–46
Mechanical analysis	47–88
Implementation	89–90

Summary of Recommendations

This section (Figs. 10-1, 10-2, and 10-3) is reproduced in its entirety. It is the section most often referred to, particularly by nontechnical members of the hospital hierarchy.

Description of Recommendations

This section (Fig. 10-4) provides greater detail of the proposals listed in the summary, and follows the same numbering for ease of reference.

Thermography

As discussed in Chap. 8, thermography is an excellent tool not only in spotting thermal losses through the envelope, but also in dramatically presenting these losses. As such, it helps sell lay personnel on the need for implementing conservation recommendations. Typical thermographs with their interpretations follow. See Figs. 10-5 through 10-9.

Architectural Observations

As indicated in the following excerpts, these observations relate to both suggestions for new construction and maintenance operations (see Figs. 10-10 and 10-11). The list includes small items which are self-explanatory and do not require further engineering analysis.

Page 1

Basis of Recommendations

The following pages describe various energy conservation recommendations, and the anticipated savings they will generate. Our calculations represent reasonable expectations of what can be anticipated by implementation of them. Note that many calculations relate to the weather and as such vary from year to year. Our estimates are based on 10 to 14 year periods of weather and therefore represent an average typical year. We have tended towards the conservative side on most of the calculations to prevent an overly optimistic assessment of the savings; thus we have used present fuel prices in calculating savings, rather than future fuel prices which will undoubtedly be higher.

With respect to implementation of savings we have included construction cost for automatic operation rather than manual. We have found manual shutdowns to be unsatisfactory since they are often neglected after some time. Construction costs for implementation are estimated costs. It is expected that the owner will get final costs from a contractor at the time of implementation. Some of the items can be simply installed directly with a contractor. Others should be undertaken in conjunction with professional advice. And still others may be done by the maintenance staff.

Your fuel and energy costs for the 12 months ending in November 1976 were:

Oil	$85,389.
Electricity	113,498.
Gas (Approx.)	3,000.
Total	$201,887.

Anticipated annual savings from specific recommendations with payback periods of 5 years or less total $36,226. This is exclusive of additional savings anticipated from better sealing of the building envelope as a result of recommendations described in SECTION 3 THERMOGRAPHY.

FIGS. 10-1–10-3 Summary of recommendations. *(Long Beach Memorial Hospital Conservation Study.)*

Page 2

Summary of Recommendations *

Proposal Number		Cost to Implement $	Annual Fuel Savings $	Annual Electrical Savings $	Total Savings $	Payback Years
1	Shutdown Laundry #-9, E-2, Ironer Fan	1600	1385	563	1948	0.82
2	Shutdown Kitchen HV-1, E-1	1000	2770	863	3633	0.27
3	Shutdown Lobby Cafeteria AC-1	1500	672	1416	2088	0.71
4	Shutdown Summer only - Operating Room Condensing Unit	0	[1](1642)	2297	655	0
5	Shutdown-Auditorium Unit	1000	92	234	326	3.06
6	Economizer Cycle Lobby Cafeteria, AC-1	6500	3151	932	4083	1.59
7	Economizer Cycle Operating Rooms, ICU, AC-4	7062	1230	574	1804	3.91
8	Recovery of Heat From Condenser Water	12400	914		914	13.57
9	Comparison of Electric vs. Steam for New 2nd Floor Alteration	3000	830	[1](1554)	830	3.61
10	Close Louvered Skylights in East Bldg. Stairwell	300	125		125	2.4
11	Use of Storm Sash	4000	705		705	5.06
12	Weather Stripping of Doors	1500	95		95	13.79
13	Use of Alarms for Exterior Doors	900	156		156	5.77
14	Use of External Awnings to Control Sun Load	3300		97	97	34.02
15	Discussion to West AC Unit Controls & Heat Exchangers				SEE DISCUSSION	
16	Miscellaneous Items (Mechanical)				SEE DISCUSSION	

(1) Negative fuel cost.

(Continued)

FIG. 10-2

Page 3

Summary of Recommendations * (Continued)

Proposal Number		Cost to Implement $	Annual Savings $	Payback Years	Comments
17	Replace fluorescents with "watt saver" tubes	175.(neg)	3840	Immed.	
18	Replacement of Incandescent Lighting (Totals with less than 1 year payback).→	9810 (1060)	5615 (1867)	1.75 (.56)	
19	Reduce West Wing Corridor Lighting by 50%.	Neg.	6572	Immed.	
20	Reduce Lighting in O.R.'s	Neg.	327	Immed.	
21	Reduce Lighting in Main Lobby	Neg.	1230	Immed.	
22	Reduce Lighting in East Patient Rooms. Remove lamp from wall fixture.	Neg.	422	Immed.	
23	Building Electric Load Profile				See Discussion
24	Minor Improvements (Electrical)				See Discussion
25	General Notes (Electrical)				See Discussion

*NOTE: Grouping of above proposals by length of paybacks, produces the following categories:

1. Paybacks of less than one year: these involve initial expenditures of $4,985 to produce an annual saving of $22,582 (a return on investment of 453%, even without allowing for future escalation of fuel prices).

2. Items with paybacks of between one to five years: these involve expenditures of $32,572 to produce an annual saving of $13,644 (a return of 42%).

3. Miscellaneous items not fitting into either of the above categories involving anticipated annual savings of between 3 to 5 thousand dollars.

FIG. 10-3

Page 4

Description of Recommendations

1 Shutdown of HV-2 (Laundry Supply), E.9(Laundry Exhaust) E-2(Laundry Exhaust) and Ironer Fan (West Wing Bldg.)

These are fan systems that are presently run continuously (24 hours a day) in the basement level. These units can be shut down for approximately 12 hours a day when not in use. This saves both electricity to run motors and in the case of HV-2 also the heating of ventilation air. We recommend this be done with a common time clock, interlocks and a night thermostat. HV-2 unit should have a return air grille and dampers for night operation. The above has been included in construction costs.

2 Shutdown of HV-1 and E-1 Kitchen Supply and Exhaust Units (West Wing Bldg.)

These fan systems for the kitchen presently run continuously. These units can be shutdown at least 12 hours a day saving electricity and heating of ventilation. A time clock and interlocks would be required.

3 Shutdown of AC-1 Cafeteria, Main Lobby unit (West Wing Bldg.)

This AC unit serves areas that could be shut down at night. Basically, the main lobby is closed from 8:30 PM until 7 AM in the morning except for occasional use by relatives of emergency medical patients. Similarly the cafeteria which is on the same system is shut down for the same hours. If these areas are occasionally used by a few personnel, we don't anticipate a problem with temperature or ventilation. If it is a problem, the time clock can be reset for shorter off hours. The systems should have a night stat control to occasionally cycle fan to maintain temperature. Construction cost includes night stat, time clock, wiring etc. This can be shut down manually on a trial basis before installing controls.

4 Summer Shutdown of Auxiliary Operating Room Condensing Unit

This unit as we understand was added to satisfy two needs. First to supplement an insufficient peak summer cooling load on the central plant and second to provide cooling during those warm winter days when the central system is shut down. Our objective here is to eliminate the use of this unit during the summer season only. This is because the electrical charges for running this equipment are much greater than the cost of steam for the main chiller plant. The reason for this is that Long Beach Hospital is charged (for electricity).

FIG. 10-4 Description of recommendations. *(Long Beach Memorial Hospital Conservation Study.)*

Page 17

Interpreting Thermographs

A thermogram is a "heat picture" of an object. Unlike ordinary photographic equipment that responds to visible light, the infrared camera responds only to invisible heat radiation. The resulting image produced on the display unit resembles the object in black and white with shades of gray. The whiter an object appears, the hotter it is. Conversely, the darker the cooler.

The sensitivity of the thermographic equipment is such that a temperature difference of as little of 0.2 degrees C. at 30 degrees C is sensed. By using an isotherm function, the actual surface temperature of an object can be measured.

An isotherm is a line or surface joining points of equal temperature, and the isotherms appear as white spots superimposed on the thermogram. As the isotherm adjustment is varied, the isotherms move to other areas on the surface of an object which are at a different temperature. For a particular temperature setting, the isotherms indicate areas on the surface of the object which are at the same temperature.

The small number at the top of the thermogram indicates the relative temperature differences between objects that are being scanned. For example, if the number is 5, objects appearing black are at or near ambient temperature and objects appearing white are 5 isotherm units above ambient. The shades of gray represent temperatures within the 5 units isotherm span.

If an object appears to be glowing "white hot," it is hotter than the maximum for the particular range. When a thermogram is taken, the equipment operator will select the appropriate range, thus avoiding the out-of-range problem. Therefore, by noting the range number, the temperatures of various objects can be estimated and where necessary, the exact temperatures can be measured using the isotherm function.

When inspecting objects whose normal surface temperatures are at or close to ambient the isotherm units vs. degrees C. relationship is approximately linear. This almost linear relationship is useful for quick estimation of temperatures that are in the 10-30 degree C. range. At higher base temperatures the relationship between the isotherm unit and degrees centigrade becomes too non-linear for accurate approximations and one must refer to calibration curves.

FIGS. 10-5–10-9 Thermography. *(Long Beach Memorial Hospital Conservation Study.)*

Page 18

Section P

Long Beach Memorial Hospital
North Wall

Ambient Temp. = 2°C

January 4, 1977
1400 Hours

S U M M A R Y

The average overall wall surface temperature of the North wall varied between 0.5°C and 1.0°C above ambient.

The major heat loss areas were found at the following locations:

Page Number	Ref. No.	Description
P.2	N.6.	The East elevator shaft showed an overall wall temperature of 3°C to 4°C above ambient.
P.3	N.8.	The Personnel Office, a conference extension, showed a large loss on the wall surface of the main building over and adjacent to the extension of 2°C to 3°C above ambient.
P.4	N.12.	The top level of the joint between the end of old wing and the beginning of the new wing showing a 4°C heat loss above ambient on top corner of new wing.
P.5	N.13.	The East side of the new wing showing heat losses around entry door, roof joint, and around entry through steel door to basement. Also note heat loss on old wing wall surface (left).

FIG. 10-6

Page 18A

Section P

(Summary, continued)

Page Number	Ref. No.	Description
P.6	N.16.	Concrete spandrel on new wing shows a 4°C heat loss above ambient. This loss was recurrent around entire new wing.

The less critical heat loss areas were those found as recurrent, low-level losses being 1.5°C to 2.0°C over ambient. As shown in thermogram, N.2. and N.5., the entire wall showed a patterned heat loss between floor levels and windows. Other areas of less critical heat loss can be found in the text, along with their respective locations.

FIG. 10-6 *(continued)*

Page 20

- P.2. -

Long Beach Memorial Hospital
North Wall (continued)

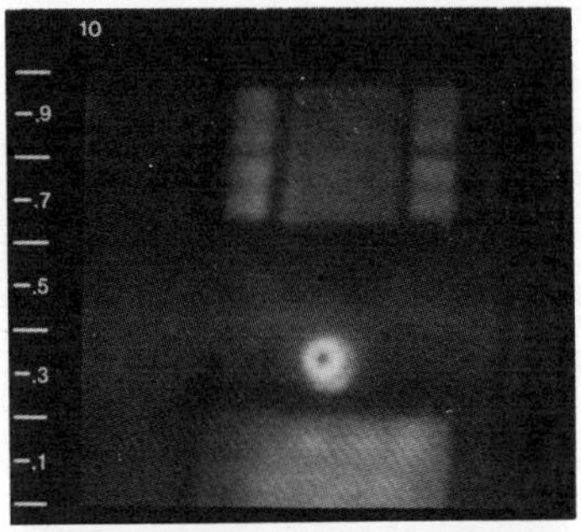

N.3. - 1st level, large picture window, administrator's office.
Note; Heat loss through exhaust fan above window. Heat loss less through middle pane of upper window. Perhaps automatic sealing should be installed when fan is not in use.

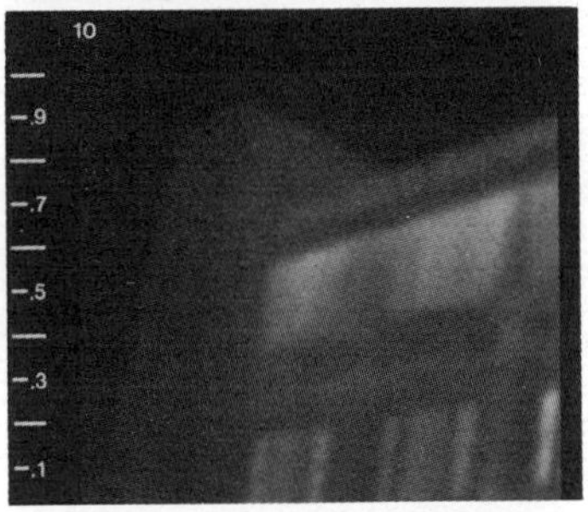

N.4. - Typical 3rd level N.E. end representative of N.E. corner of building. The heat loss area between the floor levels and building sections (vertical) can be seen as per N.1, N.2, N.3. The rest of the wall area has very minor heat loss, effectively at ambient temperature.

N.6. - East end elevator shaft showing old chimney not in use, large heat loss through windows of shaft and through the walls of the old chimney apparently not blocked off. Overall wall surface is 3°-$4^{o}C$. above ambient.

FIG. 10-7

Page 21

- P.3. -

Long Beach Memorial Hospital
North Wall (continued)

N.7. - Personnel office and conference room. No significant heat loss other than through the windows.

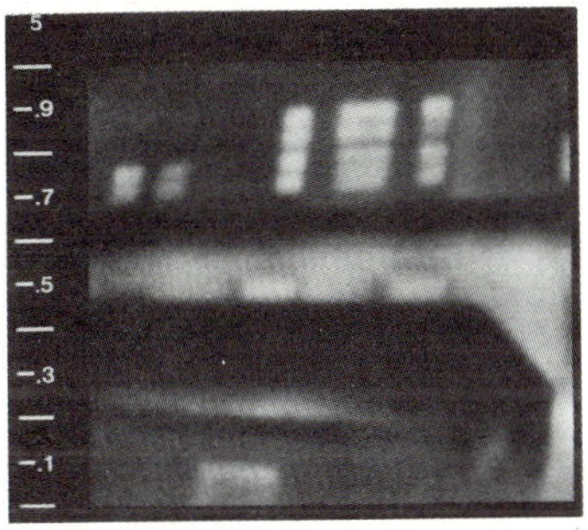

N.8. - Personnel office and conference extension, N.W. side. No significant heat loss other than small area above corner window. However, note the large loss on wall of main building over and adjacent to extension. Wall temp. = 2° to 3^oC. above ambient.

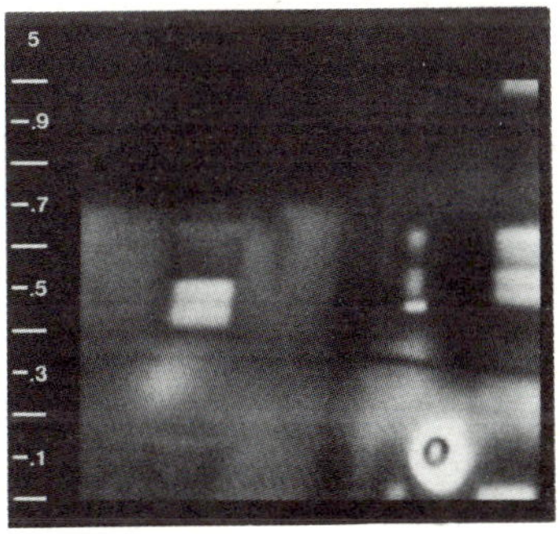

N.9. - Corridor, 1st floor nurses' station shows heat loss of 1° to 2^oC. above ambient on wall surface.
Note: Exhaust fan heat loss.

FIG. 10-8

Page 23

- P.5 -

Long Beach Memorial Hospital
North Wall (continued)

N.13.- East side of new wing showing heat loss around entry door and roof joint adjacent. Also heat loss area around steel door to basement. Old wing (left) has high heat loss.

N.14.- East end of lab, showing loss around window.

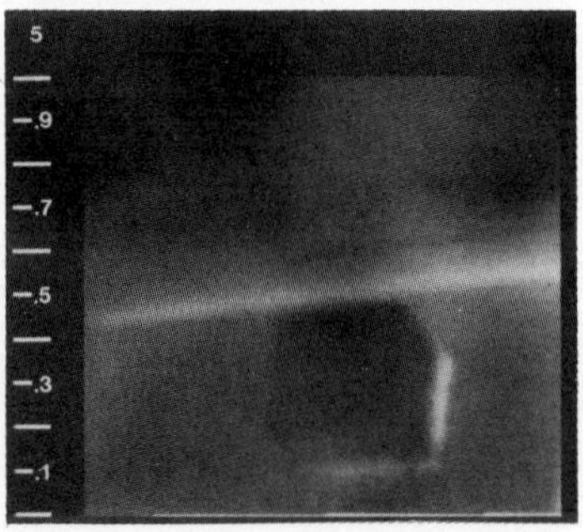

N.15.- Air Conditioner - East end of lab showing heat loss around edges indicating a poor installation - needs to be recaulked.

FIG. 10-9

Page 40

A-1 INTERIOR SPACE UTILIZATION

1. Locate wall hangings and furniture away from Supply and return grills.
2. Close off unused areas.
3. Night operations - close down Electrical, HVAC for business areas.
4. Outpatient Clinic - east door, vestibule
5. Hallway between Aud. & O.P. clinic.

A-2 INFILTRATION

1. Cellar windows and louvers original building.
2. Louver space at cellar under entry.
3. Inadvertent openings in Building's Envelope.
4. Air Curtain at Garbage doors. (McGuire-Pendador)
5. Caulking openings for conduits, pipes.
6. Door to eleva. machine area - heat flue.
7. Communicating stair exit to north window and door.
8. East stair tower - weatherstrip doors. Door to roof - skylight with 4 louvers.
9. Wind driven ventilators.
10. Southside weatherstripping at doors throughout.
11. Caulking electrical and plywood panels re exhausts.
12. Louver in door.
13. Leaky louvers.

A-3 VENTILATION

1. Cellar Morgonstern Pavillion - AC Blowing
2. West wing stair hall - skylight has fusible links operating skylight without fixed louvers.

A-4 TRANSMISSION/INSULATION

1. Ceiling over emergency - re floor over
2. Add blown insulation in walls, ceilings
3. Add roof insulation in portions of building - payback? over twenty years old.
4. Block off or thermally seal unused windows.

A-5 GLAZING

1. Broken windows.
2. Replace weatherstripping. Weatherstrip operable sash if crack is evident.
3. Conference Room - single glazing.
4. Storm window frames in place - storm windows available, other missing. Storm windows not being used because screens are stuck.
5. Other cases - lower storm is missing.
6. West wing; picture window plus sides.

FIGS. 10-10–10-11 Architectural observations. *(Long Beach Memorial Hospital Conservation Study.)*

Page 41

7. Clerestory to 3rd floor.
8. % fenestration - verify glazing
9. Storm sash
10. Awning windows - screens only (east side, auditorium etc.)
11. Curtain wall at business office.
    Caulking.
12. Old section - plastic storms on north wall.

A-6 REFLECTIVITY

1. S.S. - explore tinted film.
2. Roof surface
3. Using reflective material on window side of draperies - part of long range program.
4. Repaint exterior and or roof to provide reflective surface.

A-7 WINDBREAKS

1. Expandable enclosure for ambulance bays.
2. Close in one Bay of Emergency Entrance.

A-8 OPERATIONS

1. Doors - Gasketing weatherstripping
   Signs re: closing doors.
2. Vestibule set up for Outpatient clinic
3. Garbage doors open.
4. Emergency door open.
5. Open door north side.
6. Close shading devices at night to reduce nighttime heat losses.
7. Day room - status of drapes - south window.

A-9 HVAC OBSERVATIONS

1. A.C. units operating
2. Rest rooms - first floor - louver admitting cold air (south wall)
3. Excess Cellar heat - felt in stairway
4. Is this heat loss through pipe insulation (at ceiling)?
   What is buried Pump?
   Franklin's office - generator adjoining
5. Pit for laundry waste water heat exchange.
6. Ductwork - Air leaks, taping, caulking, insulation
7. Check interior of ducts for loose insulation - other obstructions
8. Check grilles and registers for obstruction.
9. Laundry.
10. Kitchen.
11. General: Utilization of excess heat.

FIG. 10-11

Mechanical Analysis

This section (Figs. 10-12 through 10-15) provides the engineering backup for the recommendations previously listed in the Summary of Recommendations and the Description of Recommendations. Numbering is keyed to those sections.

GOLDMAN SOKOLOW COPELAND CONSULTING ENGINEERS 212 - LE 2-9512

Page 47

HV-2, E-9, E-2 and Ironer Fan Shutdown Savings
Units Now Running 24-Hours a Day. Only Required to Operate 12-Hours a Day.

Laundry Units HV-2, E-9, E-2 and Ironer Fan

	BHP		HRS. OFF
S-6	2.8	x	12
E-9	1.5	x	12
E-2	1.0	x	12

$$\frac{63.6 \times 365 \text{ day/yr.}}{1.3410} = 17{,}310 \text{ KW hr./yr.}$$

Savings on Electric = 17,310 x .0325 = $563

Fuel Oil Savings (Using Modified Degree Day Procedure)

HV-2 CFM = 7200
Q_{ca} = 7200 x 1.08 x 70 = 544,320 BTU/hr.
D = 5219
ΔT = 70 - 0 = 70°F
n = .75
v = 150,000 BTU/gal.
C_D = .71
C_F = 1.56

$$\therefore E = \left(\frac{Q_{DA} \times D \times 12}{\Delta T \times n \times v}\right) C_D \times C_F$$

$$= \left(\frac{544{,}320 \times 5219 \times 12}{70 \times .75 \times 150{,}000}\right) .71 \times 1.56 = 4328.85 \text{ Gals.}$$

Savings on Fuel = 4328.85 x .32 = $1385.23

Construction Costs $1600

Interlocks, Common Timeclock for Existing Units, Night Cycle Thermostat, Return Air Duct Connection and Grilles, Timeclocks for New Units Recently Added Plus Wiring.

Appendix - 1

PROJECT Long Beach Hospital Date ____ No. ____

FIGS. 10-12–10-15 Mechanical analysis. *(Long Beach Memorial Hospital Conservation Study.)*

GOLDMAN SOKOLOW COPELAND CONSULTING ENGINEERS 212 - LE 2-9512

Page 48

Shutdown of Kitchen Units HV-1 and E-1

HV-1 supply unit and E-1 presently running 24 hours a day according to maintenance personnel. Only required to operate from 6 a.m. to 6 p.m. Therefore units could be off for 12 hours at night.

Electrical Savings

	BHP		HRS. OFF
E-1	3.58	x	12
HV-1	4.55	x	12

$$\frac{97.56 \times 365 \text{ days/yr.}}{1.3410} = 26{,}554.0 \text{ KW hr./yr.}$$

Savings on Electric = 26,554.0 x 0.0325 = $863

Fuel Oil Savings (Using Modified Degree Day Procedure)

HV-1 CFM = 1300

Q_{OA} = 1300 x 1.08 x 70 = 952,800 BTU/HR.

D = 5219

$\Delta T = 70 - 0 = 70^{o}F$

n = .75

v = 150,000 BTU/GAL

CD = .71

CF = 1.56

$$\therefore E = \left(\frac{Q_{OA} \times D \times 12}{\Delta T \times n \times v}\right) CD \times CF$$

$$= \left(\frac{982{,}800 \times 5219 \times 12}{70 \times .75 \times 150{,}000}\right) .71 \times 1.56 = 8656.97 \text{ GALS.}$$

Savings of Fuel = 8656.97 x .32 = $2770.23

Construction Costs $1,000

Interlocks, Common Timeclock for Existing Units.

Appendix - 2

PROJECT Long Beach Hospital Date ______ No. ______

FIG. 10-13

GOLDMAN SOKOLOW COPELAND CONSULTING ENGINEERS 212 - LE 2-9512

Page 50

Analyze Savings in Tonnage With Respect to Capacity of Separate 50 Ton Operating Room Unit

Check Tonnage Reduction

AC-1

At peak conditions 95°/75° from appendix A-2

Enthalpy under present operation	107.23 BTU/FT^2
Enthalpy under proposed operation	53.32 BTU/FT^2
	53.91

Thus Savings in Tons

$$\frac{53.91\ BTU/FT^2 \times 4602\ FT^2}{12{,}000\ BTU/TON} = \underline{20.66}\ TONS$$

AC-4

At peak conditions 95°/75° from appendix A-3

Enthalpy (present operation)	82.90	BTU/FT^2
Enthalpy (proposed operation)	54.65	
	28.25	BTU/FT^2

Thus Savings in Tons

$$\frac{28.25\ BTU/FT^2 \times 3643\ FT^2}{12{,}000\ BTU/TON} = \underline{8.57}\ TONS$$

In addition we observe that heat wheel has been inoperative for some years which means we must add an additional penalty for this increase tonnage.

From Appendix:

$$\text{Savings in Tonnage} = \frac{(40.2 - 14.07)\ BTU/FT^2 \times 3643\ FT^2}{12{,}000\ BTU/TON} = \underline{7.93}\ TONS$$

Total Savings in Peak Tons = 37.16 Tons

The above indicates there would be adequate tonnage in the existing absorption unit if the above changes are implemented. Thus analyzing the additional cooling required for the summer season as follows:

Appendix - 5

PROJECT Long Beach Hospital Date ________ No. ________

FIG. 10-14

GOLDMAN SOKOLOW COPELAND CONSULTING ENGINEERS 212 - LE 2-9512

Using AC-4 Calculations (Appendix B-2A)

Cost of Electricity to use Separate 50 Ton Chiller

$$\frac{\text{Total BTU/FT}^2 \times 1\ \text{KW}}{12{,}000\ \text{BTU/HR.}} \times .0325\ \$/\text{KW.HR} = \text{Cost of Present Operation}$$

$$\frac{89{,}040 \times 3643 \times 1}{12{,}000} \times .0325 = \$878$$

Demand

Use 40 tons at 1 KW per ton = 40 KW
Assuming this unit contributes to the peak demand which is chargeable for 12 months.

40 KW x .85 x 12 months x 3.48 $/KW = $1419

Cost of Steam When Using Steam Absorption to Handle Load

$$\frac{89{,}040\ \text{BTU/FT}^2 \times 3648\ \text{FT}^2 \times 20{,}000\ \text{BTU/HR.}}{12{,}000\ \text{BTU/TON-hr.} \times 150{,}000\ \text{BTU/GAL OIL} \times .07\ \text{eff.}} \times .32\ \$/\text{gal.} = \$1642$$

Appendix - 4

PROJECT Long Beach General Hospital Date ______ No. ______

FIG. 10-15

Implementation

This section (Figs. 10-16 and 10-17) is designed to help administrative personnel evaluate the reommendations of the report and to facilitate a decision to begin implementation.

The architect should be prepared to spend time with the client in the selection and scheduling adopted for implementing recommendations. Some of these will be targeted as part of annual operations, while others will require special financing. These recommendations should first be discussed with the building's operating engineer and then be presented to the administrative staff as a plan of action that enjoys the support of both in-house staff and the outside consultant.

Page 89

SELECTING PROPOSALS FOR IMPLEMENTATION:

Conservation options with short payback periods (less than 5 years) are more likely to be adopted than those whose benefits must accumulate over many years in the future. Short payback options produce higher rates of return on investment, faster return of capital for reuse, and comparative freedom from the hazards of error implicit in the long range forecasting required by extended life cycle benefit analysis.

At a time when fuel prices continue to escalate, however, we must emphasize that the payback periods referred to in this report were based on current fuel prices. As these prices continue to escalate, the payback periods will become shorter and make the investment even more attractive.

Traditionally many owners and institutions have thought in terms of a six to eight year payback on their money. This is entirely proper if one considers no fuel cost escalation. Plot a 10% or 15% interest rate at 0% fuel escalation on the curves on the following Fig. 10-17 and it works out to from six to eight years. However energy cost has been increasing and it will continue to increase. Thus if we plot an interest cost of 10% versus a 10% fuel increase on Fig. 10-17, the justifiable payback period becomes 20 years. Or putting it another way it is now economical to spend up to 20 times the present annual savings and still have an economical investment.

The point of this discourse is not necessarily to encourage long term investments since that is a complex issue which involves an organization's cash flow and other factors. However we think owners should be aware of the impact of fuel increases on energy saving opportunities and that we must revise our traditional thinking in this era of volatile fuel prices.

FIGS. 10-16–10-17 Implementation. *(Long Beach Memorial Hospital Conservation Study.)*

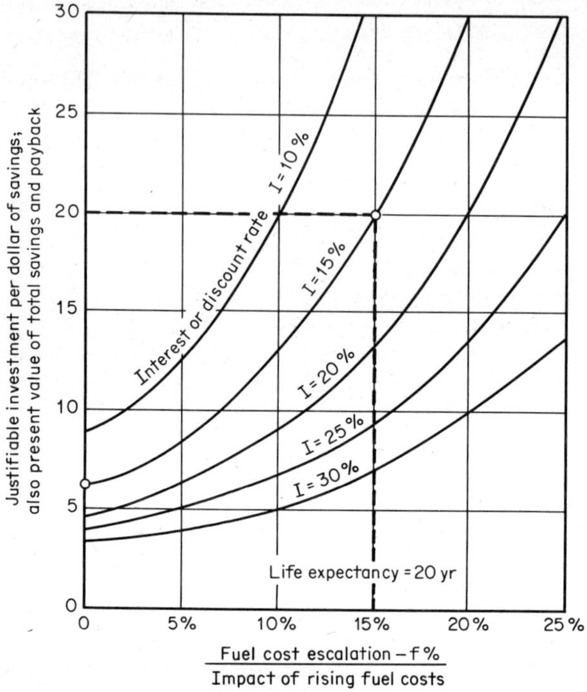

Where payback = $\frac{\text{present value of investment}}{\text{present savings}}$

References:

1. D. Carl Lehman, "Evaluating energy-saving measures," ASHRAE Journal, June 1976.
2. E. Paul Degarmo, Engineering Economy, Macmillan, 1960.

FIG. 10-17

Chapter 11

Conservation for Specific Building Types

Previous chapters have described the general energy characteristics of all buildings. This chapter supplements this general information by describing a number of specific building types:

- Office buildings
- Educational facilities
- Multifamily housing
- Hospitals
- Industrial plants
- Retail stores
- Food service establishments
- Gasoline service stations

These references apply not only to the particular type of building described, but also to other types of structures that resemble (in part or in whole) the buildings described in this chapter. Thus, nursing homes can use much of the information provided for hospitals and multifamily housing, while hotels have sections similar to multifamily housing, food service establishments, and retail stores.

OFFICE BUILDINGS

As stated in Chap. 3, the General Services Administration has estimated that energy consumption in office buildings is used as follows:

Lighting	25%
Heating	35%
Air conditioning	35%
Other	5%
Total	100%

These quantities are the result of a design process which, in the case of most of our existing buildings, was based on the presumption of maximum occupancy, people performing the most difficult tasks everywhere, and excessive metabolic functions. In analyzing an existing office building, a survey should be made to establish the actual occupancy and tasks, on a space-by-space basis, and light and ventilation provided on a realistic basis. The redefinition of these programmatic requirements is a major responsibility of the architect.

The quality of the space, as well as its adequacy for the performance of the work that takes place there, can usually be provided at substantially lowered energy levels. The collateral effect of one decision on another is obvious. Reduced lighting levels will also reduce cooling loads. At the same time they will shift heating loads from the electric system (the lights) to the fossil-fuel system. The reduction in ventilation requirements to match actual occupancy, rather than assumed maximum occupancy, and the provision of controls that will activate the ventilation system only when the space is actually occupied will result in lower fan use and reduced amounts of air to be heated and cooled.

Lighting and ventilation (including fan power and treating of makeup air) account for the majority of energy consumed by the office building. Tall office towers are visible around the country, shining brightly, often well into the night or perhaps all night, standing magnificently and opulently. They have succeeded all too well in attracting attention to themselves for the art forms they are, but it will probably be many years before we again see art so wasteful of energy.

Lighting reductions are primarily obtainable by:

1 Switching from overall lighting to task lighting wherever possible, i.e., lighting the task where it is located rather than providing overall lighting to take care of any contingency.

2 Reducing the intensity of lighting from the (erroneously) inflated standards of the past to more realistic levels. Provide illumination for office tasks of 50 fc, nominal, unless there is specific justification for a higher level. Provide the surrounding areas where only occasional tasks might be performed, or which are used for task circulation purposes, with illumination levels of between 20 to 30 fc. Provide not more than 10 fc in corridors and other public areas. It is believed that these recommendations can be met with electrical demands of 2 W/ft^2 for modifications to existing space and 1.5 W/ft^2 for new or extensively remodeled existing space.

Ventilation reductions can be achieved by:

1 Using realistic levels through code variances. The amount of oxygen required to support physical needs can usually be supplied by 1 to 2 cubic feet per minute (ft^3/min, or CFM) per capita. For usual control of odors and very light smoking 1 to 4 ft^3/min per capita is ample. Systems should be designed, therefore, not only to satisfy code levels when operating at maximum capacity but also to operate at realistic levels through simple switching arrangements.

2 Timing ventilation to correspond to actual occupancy. It is particularly wasteful when spaces are ventilated on weekends and at night when the building is sparsely populated or unoccupied.

3 Transferring energy from exhaust air to incoming air by the use of appropriate heat exchangers. This can result in sensible and/or latent heat savings of 50 to 60 percent.

EDUCATION FACILITES

The opportunities for energy conservation in our schools are well illustrated by the following activities taken from the experiences of Yale and Purdue. See Figs. 11-1 and 11-2.

MULTI-FAMILY HOUSING

Table 11-1 indicates typical targets for conservation measures. The recommendations are as follows:

1 *Replace shower heads and sink aerators.* Conventional shower heads have flow rates of 5, 7, or as much as 10 gal/min; sink faucets average 4 gal/min. Significant fuel savings are possible by reducing the flow rate of hot water by replacing all shower heads with low-flow types and installing low-flow aerators on kitchen faucets.

2 *Reduce hot-water temperature.* Excessive hot-water temperatures are both costly and unnecessary for normal apartment use. Substantial savings are possible by lowering water temperatures to more realistic levels. Hot-water temperatures should be set at 125°F (except for buildings with installed dishwashers, where 145°F should be held).

3 *Install timer on hot-water circulating pump.* Many apartment buildings use a circulating pump to bring hot water to the taps. Very often these pumps run continuously. Circulation of hot water during low-demand periods wastes heat and should be limited by installing a 24-h clock to shut off the circulating pump during low-demand periods.

4 *Relamp public areas.* Where lights burn 24 hours a day (as in interior public halls and exit stairways) change lamping from incandescent to fluorescent. In service areas where lighting is needed intermittently (meter rooms, incinerator closets, etc.) install timer switches as well. Replace incandescent

Table 11-1 Savings summary

	Recommendation	Cost to Implement	Payback Period (Yr)	First-Year Energy Cost Reduction	10-Yr Savings
	Savings on Hot Water:				
1	Replace shower heads and sink aerators.	$ 1,094	0.1	$ 9,570	$147,818
2	Reduce hot-water temperature.	No cost	Immediate	7,837	121,946
3	Install timer on hot-water circulating pump.	16	0.3	61	911
	Savings on Lighting:				
4	Relamp apartments.	1,623	0.6	3,215	39,632
	Relamp common areas.	98	0.1	1,886	29,249
	Savings on Heating and Cooling:				
5	Correct overheating.	39,000	0.8	48,583	715,616
6	Limit temperature.	5,465	0.8	6,522	96,019
7	Set back temperature (1).	6,000	1.1	5,558	80,484
	Set back temperature (2).	3,000	2.7	971	16,030
8	Improve and maintain boiler efficiency.	100	0.8	225	2,787
9	Install storm windows.	7,700	3.8	1,671	23,668
10	Insulate pipes and tanks.	200	2.9	60	921
11	Add roof insulation.	3,456	2.6	1,143	18,945
	Other Potential Savings:				
	Repair water leaks.				
	Install time switches in rooms which are used irregularly.				
	Minimize energy waste in vacant apartments.				
	Shut off pilots in gas furnaces.				
	Caulk and weatherstrip doors and windows.				

Note: The first-year energy cost reduction figure is the *gross* amount saved—the "off the top" savings on your utilities during the first year. The 10-yr savings figure is the projected *net* amount saved over a 10-yr period, before taxes. It takes into account the initial implementation expense, annual maintenance costs (if any), estimated rate of inflation, and anticipated increases in energy costs.

Source: "Energy Cost Reduction for Apartment Owners and Managers," Institute of Real Estate Management, 1977.

exterior lighting fixtures with mercury or sodium types which have much higher efficiency.

5 *Correct overheating.* When an apartment building is overheated either in whole or in part, fuel is wasted. The methods used to correct the heating system differ depending on the nature of the heating system, including its zoning and control devices. A heating contractor or engineer should be retained for this phase of the work.

6 *Limit temperature.* Thermostats set upper limits for heating and lower limits for air conditioning to prevent excess heating or cooling. In some communities, local ordinances set these limits; in others, this is a matter of landlord-tenant relationship only. In either case, the actual settings should be checked periodically to verify that the controls are set and are operating properly.

Yale University *New Haven, Connecticut 06510*

Office of the President

November 28, 1973

MEMORANDUM

TO: The Yale Community

FROM: Kingman Brewster, Jr.

RE: Energy Conservation

The officers of the University, together with the Dean of Yale College, the Acting Dean of the Graduate School and the Chairman of the Council of Masters have considered a variety of energy conservation measures for Yale University. The purpose of this communication is to inform the University Community of Yale's present position, to announce further conservation measures, and to initiate contingency planning in case of severe fuel shortages.

The Director of Operations has assured me that Yale's present fuel supply position is stable, although possible state or federal allocation plans could cause Yale a fuel shortage. In view of the uncertainty of fuel supply and governmental policy with respect to rationing, it seems prudent to undertake voluntary energy conservation measures throughout the University. We have considered and rejected a calendar modification which would have shut Yale down until the first week in February. However, we believe the national energy crisis is serious enough to require other measures outlined below which will entail some inconvenience to members of the Yale Community.

Additional energy savings will not be easy because Yale has already initiated an energy conservation program which has eliminated many of the obvious sources of energy waste. The cumulative effect of such measures, including better control of mechanical systems, lowering of temperatures, etc., yielded $300,000 of savings in 1972/73 alone (about 8% energy reduction). Yale's energy conservation efforts are complicated by the concentration of consumption in relatively few buildings, mainly in the Science Complex and the Medical School, where sensitive on-going experiments require constant climate control. For example, the Kline Biology Tower and the Sterling Chemistry Lab alone equal the annual energy consumption of all of Yale's residential colleges. Consequently, there are not many obvious ways to save significant amounts of energy short of a near total suspension of major sectors of the University's activity. There are, however, several less drastic measures which, in the aggregate, will enable Yale to achieve at least a 5% energy consumption reduction at once.

1. There will be a further reduction in the average room temperature in Yale buildings. A target of 68° has been set. The policy of reducing heat to buildings at the end of normal working hours and during weekends will remain in effect. Yale's central heating system has antiquated controls, which makes it difficult to achieve acceptable temperatures at the extremity of a steam line without overheating locations near the source. However, some pockets of overheating can be

FIG. 11-1 Yale University energy memorandum.

7 *Set back temperature.* During nighttime hours (11 p.m. to 6 a.m.) heating temperatures can be reduced to save additional fuel, subject to the requirements of local ordinances. Types of occupancy, local competition, and rental rates should be weighed against the savings that can be achieved through reduced fuel consumption.

8 *Improve and maintain boiler efficiency.* Boilers operating at less than peak efficiency waste the fuel that fires them. Most oil burners in use today can be expected to perform at a combustion efficiency of 80 percent; gas eliminated through localized valve and radiator adjustments. A special office to handle overheating reports has been established at 20 Ashmun Street. Instances of overheating should be reported to Ext. 24146.

2. The University will suspend normal operations insofar as possible during the period December 22, 1973 through January 1, 1974. Employees whose offices are shut down would be expected to count the time off as their normal Christmas vacation. Bargaining Unit employees may elect to count the time off against floating holidays and/or available vacation days. The residential colleges and all other undergraduate dormitories will remain closed until January 12, 1974. Students will not be allowed to reside in their rooms during that period. Heat will be reduced in all buildings affected by the shutdown to 50° and hot water service will be suspended. Masters', Deans' and Resident Fellows' apartments will be heated through localized units. It is estimated that this action will save over 325,000 gallons of fuel. A building-by-building schedule will be issued later this week. However, Science and Medical School buildings containing on-going research, the Health Center, the main Library and certain essential administrative offices will remain open. The Hall of Graduate Studies, Helen Hadley Hall and the Medical School dormitories will remain open. A few dormitory spaces in Ray Tompkins House will be available for undergraduates, especially those from foreign countries, who are not able to return home for Christmas. Applications for such space should be made to the Yale College Dean's Office.

3. I have asked the Acting Director of Personnel to review the possibility of a change in working hours to take full advantage of daylight hours. The normal work schedule might be from 8:00 A.M. to 4:00 P.M., with one-half hour for lunch (8:30 A.M. to 4:30 P.M. if Daylight Savings is restored).

4. I have asked the Secretary of the University to consolidate in as few buildings as possible all public and extracurricular events occurring on weekends and in the evening.

5. I have asked that a specific energy conservation plan be developed for each building by the Engineering Section of Buildings & Grounds Service and that an energy monitor be appointed on a building-by-building basis. This plan should include considerations of temperature, schedule of use, illumination levels and tighter control of mechanical systems.

6. I have asked the Chief of the Campus Police to review external illumination for the purpose of reducing it consistent with safety requirements. Decorative lighting of buildings, such as Harkness Tower, has been eliminated. No external Christmas decorations requiring energy will be erected this year.

7. I have asked the Department of Buildings & Grounds Services to review all designs of new and renovated buildings to minimize energy consuming features of such designs. There will be a presumption against comfort air conditioning in new buildings or major renovation requests. I have asked for a review of comfort air conditioning throughout the University for the purpose of shutting down or severely restricting the operation of all units.

8. I am appointing a Faculty Energy Conservation Advisory Committee to work with the Director of Operations in exploring further possible

FIG. 11-1 *(continued)*

burners at the same. Many boilers have never been properly set up for the most efficient combustion since this must be done on-site by a qualified technician. Even boilers which once were properly tuned will degenerate over time as heat transfer surfaces get dirty, boiler draft changes, and combustion air settings get out of adjustment.

9 *Install storm windows.* Storm windows help keep heated air inside the building and will save 10 to 20 percent of heating costs. During summer conservation measures, especially in high energy consuming buildings in the Science and Medical areas. The Committee will consist of:

Charles K. Bockelman
Elias Clark
Radley H. Daly
Georges May
Merton J. Peck
Frederic M. Richards
George Veronis

Members of the Yale Community are invited to bring suggestions to the attention of members of this Committee.

The energy crisis is not likely to pass quickly or quietly from our lives. While no one can predict with certainty what fuel shortages will mean for our accustomed way of life this winter or beyond, it seems likely that more fundamental alterations in our patterns of work study and recreation are in prospect. Consequently, I have asked the Director of Operations, in consultation with the Advisory Committee, to prepare contingency plans for fuel shortages based on 10-20-30% of present consumption. The Officers and Deans are directed to explore the costs and benefits of a major calendar revision for the 1974/75 school year which would close the University from before Christmas to early February. Such a calendar change may represent the only feasible remaining source of energy savings after the measures outlined in this memorandum have been implemented.

Many of the specific proposals suggested here for energy conservation depend on the thoughtful cooperation of all members of the Yale Community. I hope the cumulative effect of those measures will spare Yale more fundamental dislocations.

KBJr:chs

FIG. 11-1 *(continued)*

months they also help retain cooled air. Because their costs result in fairly long payback periods, they tend to be installed only when their other advantages (reduction of street noise or grime penetration) provide additional tenant benefits.

10 *Insulate pipes and tanks.* Bare steam and hot-water pipes waste money and should be properly insulated.

11 *Add roof insulation.* Some roof insulation is necessary for every struc-

PURDUE UNIVERSITY
OFFICE OF THE BUSINESS MANAGER AND ASSISTANT TREASURER
WEST LAFAYETTE, INDIANA 47907

TO ALL EMPLOYEES OF PURDUE UNIVERSITY

A year ago, we asked each of you to join in a cooperative, volunteer effort to help Purdue save $180,000 in 1972-73 utility costs--funds which could be used for other programs, to pay increases, and to purchase needed supplies.

The evidence is now in and it clearly shows that you not only cooperated but went well beyond an original expectation that economies in our utility usage--water, power, lights, heat, air-conditioning, natural gas, and so on--could mean as much as a $180,000 reduction in utilities cost to be made available elsewhere in the university.

Take a bow! The combined savings in our utilities budget exceeded five percent, actually totaling $212,689! What does that mean individually? At a time when the university's operating funds are critically short, it means that many employees who received pay increases would not have gotten them because these funds otherwise would have literally gone up the stack!

And in that battle against economic and ecological waste, the amount of water saved (and thus power used to pump it out of the ground and distribute it) was a phenomenal 207,000,000 gallons, a ten percent reduction for a period in which the Physical Plant Department had expected a sharp increase.

Our power usage increased by only 20,000 kilowatt hours in 1972-73. That is a drop in the bucket, really, since the university used 129,400,000 KWH and traditionally has experienced a six to seven percent annual increase in purchase and Purdue-generated power usage.

Of course, other factors enter into the saving, but our engineers have analyzed the entire situation by several methods and now conclude that, overall, the primary reason is because people did respond to the request and did cooperate in frugal use of all of our utilities.

So, thanks to all of you, Purdue did win the battle in 1972-73--yet, the "war" goes on, and in 1973-74 we look forward to your continued cooperative effort to help save the university's resources.

Recently hundreds of time clocks were installed to assist in the most efficient use of fresh air ventilating systems, to save up to $20,000 annually. The Physical Plant staff is also conducting what eventually will be a campus-wide survey of interior lighting to reduce light to recommended levels wherever possible.

As an example of how dramatic that can be, fluorescent tubes were removed from every other fixture on the second floor of the Administrative

FIG. 11-2 Purdue University energy memorandum.

ture. The use of additional insulation is a matter for engineering calculation since it varies with the type of structure and its location.

Finally, electric metering for tenants should be on an individual basis rather than master-metered. Studies on the subject indicate that when tenants have to pay their own electric bills directly, their consumption is 30 to 40 percent less than when the charge is paid by the landlord via master metering. While individual metering is easily accomplished in new construction, its implementation in existing structures is an exercise in landlord diplomacy.

Services Building. Light levels are not only sufficient, but many persons have said the reduction has eased eye strain. The calculated saving in a year's time, if this reduction is extended to the rest of the building, is amazing. Reducing light levels to 100 foot-candles in this one building would mean:

-- A reduction in coal burned (to produce power) of 452.5 tons and a resultant reduction in sulphur dioxide pollutants of 11.8 tons!

-- A maintenance savings--new tubes, fixtures, etc.--of $2,210 and an operating utilities savings in one year of $10,182, or a total of $12,392.

It is not too difficult to imagine the large savings which we can make with little effort in other buildings across the campus.

A word about water: We pumped 207,000,000 gallons less last year than in the previous year. Cutting out unnecessary water usage saves not only our supply of clean water but dollars, too. We calculate that it costs five cents per 1,000 gallons to pump and distribute water through our campus mains. A small sum? Perhaps, but we must also get rid of that water and we pay nineteen cents per 1,000 gallons to let it pass through West Lafayette's sanitary disposal plant. That doesn't include, of course, some air conditioning cooling water on the central campus which we save to water our vast campus lawns.

This, in part, is what the university is doing to help save on utility costs. Here's what you can do:

1. Stay on sidewalks. Why? The university annually sods about 10 acres of new lawns because of construction work and similar reasons. But one unnecessary cost is the $4,000--at about ten cents per square foot--spent to "repair" lawns where pedestrian paths have been worn.

2. Protect air conditioned and heated areas from unnecessary temperature change loads. For example, doors left open between air-conditioned and nonair-conditioned areas cost about $27 per year per door. Keep such doors closed.

3. Windows left open in air-conditioned areas cost even more--$124 per year per window. On the subject of windows, window air conditioners should be set so the thermostat can operate. For example, leaving a window air conditioner on the "constant cool" mark means the conditioner runs full blast, whether it needs to or not. Set the conditioner so the thermostat works and the compressor shuts down when not needed. Figure savings at about two cents per hour. There are at least 1,700 window units installed on campus.

4. Lights in a typical Purdue classroom cost about six cents per hour to use. Turn them off when a room is not occupied.

5. Lights in a typical residence hall room cost one half cent per hour. Consider the hourly savings if the use in each of the more than 6,000 rooms was reduced one hour per day.

6. Laboratory aspirators use two gallons of water per minute. The cost of providing water to the aspirator and of disposing of the effluent

FIG. 11-2 *(continued)*

HOSPITALS

On the average, energy use in a hospital is divided among six categories:

1 *Environmental control.* Heating, ventilating, and air conditioning (HVAC) require 40 to 65 percent of all energy used in a hospital.

2 *Lighting and wall receptacles.* Hospitals operate around the clock seven is 60 cents per day per aspirator. There are more than 540 aspirators available in the Pharmacy Building alone, not to mention the others in other university facilities. Turn them off when not needed.

7. Laboratory hoods require 1,000 cubic feet of air per minute during operation. In air-conditioned areas, replacement air costs nine cents per hour per hood at design conditions to satisfy room conditions. Shut hoods off when not needed.

8. Access to controlled humidity areas should be kept to an absolute minimum. The "price" of opening a single door between a controlled humidity area and a corridor or uncontrolled area is two cents.

9. Consider the added cost of setting your room thermostat at 72 degrees rather than 75 in summer and 80 degrees versus 75 in winter: a typical classroom with the thermostat at 72 degrees summer and 80 degrees winter, as against a year-round 75 degree setting, costs $35 more.

10. Scheduling anticipated large electric power loads during off-peak loan periods can save hundreds of dollars annually. Example: A 25 h.p. electric motor run for one 30-minute period during the peak load periods costs $41.60 more for energy per month than if run during an off-peak period. The peak load periods here occur between 8:30 and 11:30 a.m. and from 1:30 to 4:00 p.m. on weekdays. You should reschedule anticipated large and small electrical loads to miss peak load periods whenever possible.

If it seems that the university expends much power and water on lawn sprinkling, remember--that is one of the prices of maintaining lawns and landscaping in this climate. But remember also that the university uses air-conditioning, cooling, and condensate water whenever possible for lawn sprinkling--water that otherwise would be wasted down the drain and which could create thermal pollution in the university's drainage pond on the south edge of the Agriculture Campus. Such sprinkling is considered an ecological plus.

Most of the items mentioned here would appear to be an insignificant factor in a total utilities budget of $4.1 million when considered individually. But consider them together and we estimate that the above ten measures alone could save five percent of the total utilities expenditure in 1973-74.

A goal of five percent reduction in utilities cost by extra care on the part of all of us is not unattainable and the effort has a real dollar payoff, as we so vividly learned in the last year.

So close a door, turn off a light, shut down equipment when not in use, leave that thermostat at a constant setting. It will save <u>your</u> budget dollars. And there is another plus:

Power used at Purdue, like power used anywhere, is creating pollution for someone, somehow, someplace on this planet. When you save power, you are literally saving your environment!

L. J. Freehafer
Vice President and Treasurer

FIG. 11-2 *(continued)*

days a week. Lighting and wall receptacles account for 10 to 20 percent of all energy use.

3 *Laundry.* An in-house laundry can use 8 to 15 percent of the energy total.

4 *Food service.* Kitchen operations account for 5 to 10 percent of all energy in a hospital.

5 *Medical equipment.* This equipment uses about 3 to 5 percent of all energy.

6 *Sterilization and incineration*. These activities account for about 2 percent of the energy use.

Experience suggests that as energy costs rise to 3 to 5 percent of a hospital's total budget, energy conservation is sensed as an idea whose time has come. Architects familiar with hospital work know too well the amount of red tape and committee consultations this work requires. Energy conservation is no exception, and it results in yet another committee, for experience further suggests that no sustained, hospitalwide effort can succeed unless it is supported by an energy management committee.

In effect, this committee builds a framework within which detailed, day-to-day tasks of energy management take place. The committee is a tool, not an end in itself. It serves as part of top management to provide and coordinate two-way dialogue with energy end users throughout the hospital as well as evaluate a hospitalwide impact of proposed energy saving ideas, especially safety and health care impact.

In evaluating energy-saving suggestions, the following questions should be kept in mind:

- Is the idea basically feasible? Has it been done elsewhere? In a hospital?
- Are there any clear-cut health hazards? Is there a danger of contamination?
- What is the overall impact on the quality of health care?
- Does the item fall within prudent practice and applicable regulations?
- What is the estimate of all costs to implement the idea?
- What are the estimated energy and dollar savings each year?
- Is it worth the effort, the risks?

In searching for energy-saving ideas, remember that a modern general health care complex usually includes spaces in which nonsensitive functions are housed. A teaching facility consisting primarily of classrooms (not laboratory or research stations) is essentially a schoolhouse. A lecture hall without health care or laboratory demonstration facilities is an auditorium, or theater, or movie house. A nurses' residence is a dormitory. A section of a building containing office space is an office building. A food service station, if its air pattern is not intermingled with sensitive areas, is a cafeteria. A warehouse is a warehouse. A laundry is a laundry.

In many hospitals, airflow patterns and energy distribution circuits are intermingled with sensitive areas. If it can be established that nonsensitive areas operate independently, then energy management for the system (equipment or unit) serving these areas will be essentially similar to "civilian" counterparts. Similarly, the lighting in many hospital areas may be viewed as being in an office building. If lighting can be reduced and the air conditioning modified to take advantage of the reduction, then a substantial saving is realized.

Often previously sensitive areas are converted to nonsensitive functions without modification of the HVAC system. The result can be offices served with 100 percent outside air and an HVAC system operating unnecessarily all night long.

INDUSTRIAL PLANTS

The energy consumed by industrial plants goes to space conditioning the buildings and to the industrial processes which are necessary for production. Within these two categories, there are five sometimes-overlapping approaches which suggest ways to optimize the use of energy:

- Heat confinement
- Scheduling
- Heat recovery
- Proper equipment
- Maintenance

Space Conditioning

Heat Confinement

1 Install insulation. Rapidly increasing fuel costs have made insulation cost-effective in more applications than ever before. Some companies have even found it practical to spray water on building roofs to help mitigate the effects of solar heat during warm weather.

2 Reduce unnecessary ventilation where safety is not an overriding factor.

3 Cut leakage from the building by installing proper weatherstripping and caulking and by properly enclosing loading docks.

4 Reduce conditioning in unoccupied space to a minimum.

Scheduling

1 Reduce heating or cooling during those times when the space is unoccupied. Time clock controls can be very useful here.

2 Reduce or eliminate lighting wherever and whenever possible. Time clock controls or photocell switches should be considered for exterior and decorative lighting. In many cases most indoor lighting close to the perimeter of the building can be turned out during the daytime.

Heat Recovery

1 Waste heat from burner stacks and industrial processes can be recovered and used in many cases.

2 Recycle waste materials (sawdust, oil, etc.) as supplemental fuels where possible. In many cases this represents not only a source of cheap fuel but also a reduction in pollution problems.

Proper Equipment

1 Check with equipment suppliers or qualified engineers to determine the extent to which HVAC equipment is optimally sized for the job to be performed. In many cases it is oversized, and occasionally one boiler in a multiboiler system can be eliminated by more efficient use of the others.

2 Consider the use of infrared heaters in high bay areas and warehouses. Infrared requires less fuel because it heats solid objects (personnel) in the radiation path without having to heat the air in the room.

3 Replace incandescent lighting with lamps of higher efficacy.

Maintenance

1 Emphasize boiler maintenance. Cleanliness of the heating surfaces (both on the water side and on the fire side) and maintenance of the proper air-fuel ratio at the burner are two important variables impacting on boiler efficiency.

2 Monitor stack temperature and the concentrations of carbon dioxide, oxygen, and carbon monoxide in the flue gas from combustion processes. A stack temperature that is excessively high means wasted heat. The presence of carbon monoxide in the flue gas indicates incomplete combustion. Oxygen should be kept below 2 percent, and carbon dioxide should be minimized in the range of about 8 to 12 percent, depending on the fuel used.

Industrial Processes

Heat Confinement

1 Insulate steam and hot-liquid pipes, furnaces, and other heated containers where possible. Keep openings in equipment closed and sealed, and use reflective heat shields where openings are necessary.

2 Direct or confine flames to the spot where needed. Optimize flame geometry to help accomplish this. Substitute sealed-in burners for open-hole firing.

3 Insulate water-cooled work rails or skids or construct them from refractory materials to minimize heat losses.

Scheduling

1 Avoid short runs on thermal process equipment and shut-down or idle equipment during production interruptions.

2 Schedule proper loads so that equipment can perform at optimal levels of efficiency. Survey the plant for improperly sized motors, etc.

3 Electricity can be used more efficiently if some use can be diverted to off-peak periods when voltage remains high.

Heat Recovery

1 Recycle waste as fuel and utilize waste heat (as suggested for space conditioning) to preheat boiler feed water or incoming materials which are to be heated.

2 Install recuperators on combustion equipment to preheat incoming combustion air.

3 Return steam condensate to boilers.

Proper Equipment

1 Investigate the practicality of heating liquids by immersion or submersion heating rather than underfiring.

2 Use direct-flame impingement or infrared heating for chamber-type heating where applicable.

3 Convert to higher-voltage electric systems where practical.

4 Load-shedding devices can automatically shut off less crucial machinery when voltage drops so that power is maintained on the more crucial machines.

Maintenance

1 Check and maximize the power factor in the plant. Make sure that the plant is equipped with enough properly functioning capacitors to correct the power factor.

2 Check steam systems regularly for leaks and poor insulation, and provide steam traps where needed.

3 Make a special effort to keep all heat-transfer surfaces clean.

4 Watch for hot spots on furnaces for signs of deterioration in the lining, and reline or insulate accordingly.

RETAIL STORES

Depending on the role of lighting in merchandising and the nature of the products being sold, lighting for interiors and outdoors typically consumes 40 to 60 percent of total energy costs. Heating, ventilating, and air conditioning are seasonal and show the expected climatic variations from 30 to 50 percent. If refrigeration is used (as in supermarkets), it may consume from 13-30 percent of energy costs. Hot water generally represents from 1 to 6 percent of total costs.

Targets for energy conservation include

- Interior lighting
- Outdoor lighting
- Space heating
- Air conditioning and ventilation
- General maintenance
- Warehousing
- Refrigeration

Interior Lighting

Because lighting plays such an important role in merchandising in most retail stores, it represents an area of important savings potential. Many retail establishments are overlit by modern efficiency standards, and others are inefficiently lit, in particular when display lighting is considered. Depending on the store's geographic location, lighting is probably the greatest energy expense. Possible savings can be obtained by doing the following:

- Relamp with low-energy types.
- Remove all unnecessary lamps.
- Use lower-wattage lamps.
- Control light use.
- Use more efficient sources.
- Use task/display lighting.
- Use efficient security lighting.
- Maintain (clean) lighting system.

Outdoor Lighting

Outdoor lighting is a significant energy cost in retail operations for two distinct purposes—merchandising and security or parking. For each use, different requirements may apply. Outdoor lighting for signs and display purposes is necessary from dusk to the close of operations, approximately 1500 h/yr. It is considered desirable to have good color rendition and relatively high light levels—5 to 25 fc. For purposes of parking lot lighting or security, color is not critical; a light level of 1 to 2 fc is acceptable. Security lighting from dusk to dawn is about 4380 h/yr. Other elements which affect outdoor lighting are physical layout and building site, special zoning requirements regarding light trespassing, maintenance considerations, and pole placement limitations.

Space Heating

By modifying current space heating practices as follows, meaningful savings can result without sacrificing loss of sales, creating customer discomfort, or damaging merchandise:

- Set back temperatures to 45°F during nonworking hours.
- Lower heating levels so that customers entering with their coats on feel comfortable.
- Reduce heat loss in display windows through proper insulation.

Air Conditioning and Ventilation

Savings result from taking these actions:

- Set forward air-conditioning temperature to 75°F to 78°F.
- Shut off cooling system and reduce load during nonoperating hours.
- Moderate outside air intake.

General Maintenance

Keeping any system running at its optimum operating level will save considerable energy dollars. This can be done by improving boiler efficiency, reducing infiltration losses, insulating pipes and tanks, and repairing water leaks.

Warehousing

The potential for energy cost reduction in warehouse space is quite substantial because in many retail operations, depending on the type of product, a large precentage of the total square footage is used for the storage of merchandise. Unlike the sales areas used for merchandising, warehouses do not have to be maintained at comfortable levels of lighting, heating, cooling, or ventilation. Further, the possibilities of cost reduction around loading docks and garage areas are significant because of loss of conditioned air through openings.

Refrigeration

When a retail owner requires refrigeration, it can amount to up to 30 percent of the total energy cost. An important factor in reducing operating costs is maintaining all components of the cooler unit in good condition. Insulation with weatherstripping or sealants prevents energy loss. Other elements in efficiency include control of door opening, turning off equipment at night or raising the temperature setting (this varies with the merchandise), and turning off lights inside the cooler at night.

FOOD SERVICE ESTABLISHMENTS

From the moment the food service establishment opens in the morning until the last consumer is gone, the clean-up crew has finished, and the manager turns off the lights and goes home, this type of business is a heavy consumer of energy. While most people are asleep, its freezers, refrigerators, night lights, and alarm systems continue to operate. Typical usage for food service operations is shown in Table 11-2. The four major types of energy-consuming equipment used by food service operations are food preparation and storage; lighting; heating, ventilating, and air conditioning; and sanitation.

Food Preparation and Storage

Almost half the total amount of energy required by a typical food service establishment is used to prepare and store food. Most of this energy is consumed by cooking equipment—ovens, broilers, fryers, griddles, ranges, steam cookers, and warmers. Close to 60 percent of this energy may not be used to actually cook food, but is lost to the kitchen as waste heat. Regardless of the type of equipment, the following principles should be observed in its use:

- Cook in the largest volume possible.
- Cook at the lowest temperature that still gives satisfactory results.
- Reduce excess heat loss by carefully monitoring preheat times, cooking temperatures, and maintenance checks.
- Reduce peak loading to reduce demand charges.

Table 11-2 Yearly energy use and costs

Type of Food Service Operation	kWh/yr	Natural Gas (Mcf/yr)	Million Btu/yr	Cost per year
Fast food	216,000	1,500	2,237	$ 7,320
Coffee shop	249,821	1,826	2,679	8,648
Table service	284,000	2,500	3,469	10,680
Cafeteria	473,850	4,754	6,371	18,985
Hotel/motel	852,974	5,666	8,577	28,391

Source: "Guide to Energy Conservation for Food Service," Federal Energy Administration, 1977. Sample data for 1973 were compiled by the Midwest Research Institute. They were adjusted and updated to reflect 1975 national averages by Resource Planning Associates.

Lighting

While lighting costs represent a relatively small percentage (8 to 10 percent) of a food service facility's total energy bill, lighting conservation offers immediate energy and dollar savings. The Illumination Engineering Society's (IES) recommendations for lighting intensity and lamp types are shown in Table 11-3.

Table 11-3 Food service lighting recommendations

Rooms	IES Recommended Illumination in Foot Candles	Lamp Type
Kitchen overall work surfaces	30–70	Fluorescent
Dining Room (formal) dining area foyer cloak room along walls at center	3–10	Incandescent
Dining Room (informal) dining area foyer along walls at center	15–20	Incandescent
Cafeteria, Coffee Shop at counter along walls at center	50–100	Fluorescent
Corridors, Stairs, Elevators	20	Fluorescent
Entrance Foyer	30	Fluorescent
Lobby	10	Fluorescent
Office	100	Fluorescent
Public Toilet	20–30	Fluorescent
Exterior Entrance	10	Sodium
Active Store Room	10	Fluorescent
Inactive Store Room	10	Fluroescent/ sodium
Walk-in Refrigerators	10	

Source: Federal Energy Administration, "Guide to Energy Conservation for Food Service."

Heating, Ventilating, and Air Conditioning

More than one-third of the energy consumed by a typical food service facility is spent on heating, ventilating, and air conditioning—an expense second only to the energy required for food preparation. A restaurant has special heating, cooling, and ventilating needs. Ventilation is the key to a food service establishment's HVAC system, and must be capable of filtering and removing grease and fumes from the air at rates required by local building and fire codes and by the 1971 Clean Air Act. Some recent studies have concluded that ventilation rates currently specified by fire and building codes could be reduced by 50 to 55 percent without hazard, and it is worth investigating with local officials. Heat recovery devices (heat wheels, coils, or pipes) are being used increasingly to transfer heat from the exhaust to incoming air.

Sanitation

Dishwashing and utensil sanitation are vital for the protection of both public health and a food service facility's public image. These sanitation processes alone consume about 13 percent of a restaurant's energy budget. Water-heating systems heat water to temperatures appropriate for dishwashing, food preparation, and lavatories. A water heater provides hot water to faucets in one of two ways: water can be heated to the temperature at which it will be used at the faucet, or it can be heated to a higher temperature, stored, and then tempered with cold water before it reaches the faucet. Some food service establishments maintain separate water-heating systems for kitchen (very hot water) and for lavatory use (warm water). Increasingly, various forms of heat recovery systems, either from kitchen equipment or from the air-conditioning system, are being used to preheat water and thus reduce the amount of fuel consumed for this purpose.

GASOLINE SERVICE STATIONS

Table 11-4 indicates typical targets for conservation measures. The recommendations are as follows:

1 *Relamp interior areas.* The replacement of lamps with low-energy types (mercury and sodium) and selective removal of lamps will significantly reduce lighting costs without affecting lighting quality. Consider using timers on light switches in areas where traffic is limited (as in toilets).

2 *Relamp outdoor areas.* More efficient lamp types in outdoor areas result in lower energy consumption. Where possible, change to high-pressure sodium and reduce light levels in overlit areas.

Table 11-4 Summary of savings

Recommendations	Implementation Costs	Payback (yr)	First-Year Reduction	10-yr Net Savings
1. Relamp interior areas.	$ 190	1.6	$ 114	$ 1,542
2. Relamp outdoor areas.	4,200	3.3	1,260	13,683
3. Set back heating temperatures.	120	0.5	230	3,702
4. Reduce energy use in vending machines.	50	0.5	96	1,409
5. Maintain air compressor efficiency.	0	Immediate	105	1,595
6. Improve combustion efficiency of oil burners.	100	0.5	185	1,393
7. Reduce heat loss from open garage doors.	0	Immediate	311	5,812

Note: The first-year reduction is the *gross* amount saved—the "off the top" savings on utilities during the first year. The 10-yr reduction is the projected *net* amount saved over a 10-yr period, before taxes. It takes into account the initial implementation expense, annual maintenance costs (if any), estimated 6 percent annual rate of inflation, and anticipated increases in energy costs of 9 percent per year.
Source: "Energy Cost Reduction for Gasoline Service Stations," Federal Energy Administration, 1977.

3 *Set back heating temperatures.* During unoccupied periods, a temperature of 45°F is sufficient to prevent freeze-ups. Check to make sure that thermostats are operating properly and can be set as low as 45°F; if not, replace with an automatic temperature set-back system. Daytime temperatures should be set at about 62°F.

4 *Reduce energy use in vending machines.* Operation of vending machines during nonworking hours and use of lighting on vending machines are usually unnecessary. Install an automatic timer to cycle machines off during closing hours. Remove lamps from machines.

5 *Maintain air compressor efficiency.* High air compressor settings and leaks in the hose line can account for electric energy waste. Air compressor settings should be reduced to minimum levels and air leaks repaired.

6 *Improve combustion efficiency of boilers.* Boilers operating at less than peak efficiency waste the fuel that fires them. Boilers should be set for maximum combustion and maintained for maximum efficiency. Boilers should be cleaned and adjusted by a qualified technician at least once a year.

7 *Reduce heat loss from open garage doors.* Open garage doors probably represent the largest heat loss in service areas. Reducing the length of time that garage doors remain open prevents this heat loss. Set up a regular procedure for closing garage doors manually, or install a mechanism for automatically minimizing door openings.

Chapter 12

The Single-Family House

The single-family house is a building whose energy behavior is similar in broad principle to other structures and therefore it can be analyzed following the procedures described earlier.

As a practical matter, however, it is doubtful that the architect will ever receive a fee sufficient to pay for a thorough analysis. A 10 percent fee of an annual energy consumption of $2000 is $200, hardly enough to cover the costs of assembling data, much less their analysis and the preparation of recommendations.

Accordingly, both industry and government have prepared broad guidelines describing the energy activity and recommendations applicable to large groups of homes. By referring to these recommendations and selecting those items applicable to a particular case, the architect is able to quickly put together a prescription, being careful in the process to verify that the house does conform to the handbook standard and is not an exception.

The basis for these "cookbook recipes" has to be carefully examined by the practitioner. Given two identical houses which are sited identically, the homes will behave differently if heated by two different types of systems. If one house is heated by fossil fuels, it will exhibit a much greater infiltration loss from the air being sucked in by the oxygen-consuming furnace; a solar or electrically heated home will exhibit mostly crack infiltration. The first (fossil-fuel-heated) home may spend most of its energy treating infiltrated makeup air, while the other home exhibits largely conduction losses and thus makes a better subject for increases in insulation.

On a nationwide average, residential energy is used as follows:

Space conditioning	70 percent
Hot water	20 percent
Lighting, cooking, and appliances	10 percent

SPACE CONDITIONING

The heating system can account for up to one-half of the total primary annual energy consumption of the average single-family dwelling. The air cooling system generally accounts for between 10 and 20 percent of the annual energy consumption.

The elements in the house which place the major loads on the space conditioning systems are shown in Table 12-1.

Infiltration represents the largest single factor; and too much emphasis cannot be placed on caulking, weatherstripping, storm sash, and a direct supply of makeup air to the fossil-fuel-burning boiler.

The National Association of Home Builders, in its "Thermal Performance Guidelines for One and Two Family Dwellings," states:

> The importance of reducing air infiltration cannot be overstressed. The amount of glass area and the quality of windows and doors greatly influence the amount of air infiltration. Poorly fitted windows and doors, not weatherstripped, can negate much of the benefit derived from extra insulation and other energy conserving techniques.
>
> Minimize the use of exhaust fans in cold climates and consider using those that close tight.
>
> Consider:
>
> - Sill sealer between the top of the foundation and the band joist or sill plate.
> - Sill sealer or caulking between the bottom exterior wall plate and floor sheathing or slab.

Table 12-1 Average United States residential energy consumption

	Heating Load (%)	Cooling Load (%)
Ceiling	3.7	2.3
Floor	2.2	2.4
Total window	13.6	4.1
Total door	1.4	0.4
Total wall	26.9	12.7
Infiltration load	52.2	43.0
Internal load	—	35.1
Total	100.0	100.0

Table 12-2 Air changes occurring under average conditions in residences, exclusive of air provided for ventilation*

Kind of Room	Number of Air Changes per Hour
Rooms with no windows or exterior doors	0.5
Rooms with windows or exterior doors on one side	1
Rooms with windows or exterior doors on two sides	1.5
Rooms with windows or exterior doors on three sides	2
Entrance halls	2

*For rooms with weatherstripped windows or with storm sash, use two-thirds these values.

Source: Reprinted with permission from the ASHRAE *Handbook and Product Directory, Fundamentals*, 1977.

- A 1 × 4 for the bottom wall plate rather than a 2 × 4.
- Caulking cracks around doors, windows and other openings or penetrations of the walls and at corners.
- Higher grades of caulking, they are more cost effective since labor remains constant. Latex, acrylic, butyl rubber and chloro-sulfonated polyethylene caulks are intermediate performance types while high performance categories include polysulfide, polyurethane and silicone.
- Nailing sheathing tightly to the framing.
- Replacement of wall sheathing damaged during construction.
- Ductless kitchen and bathroom fans.

In checking existing houses for infiltration, it may be helpful to move a lighted candle around the frames and sashes of windows; if the flame dances around, caulking and/or weatherstripping is needed. Try slipping a quarter under an exterior door; if it goes through easily, weatherstripping is needed.

Ceiling and Wall Losses

The figures shown in Table 12-1 indicated that ceiling and wall losses total 30.6 percent. Studies of insulation manufacturers usually quote losses from 35 to 60 percent for these same elements; ASHRAE procedures generally result in losses calculated to be between 40 and 65 percent—based on the assumed air changes shown in Table 12-2.

By contrast, in the spring of 1977, Applied Science and Engineering of Colorado went into 31 Denver-area homes with monitoring equipment to measure actual heat losses within homes; their measurements showed ceiling and wall losses totaling between 17 and 28 percent. The need for insulation relates to heat lost through the envelope, walls, and ceiling by conduction and it is obvious that a large amount of field verification still remains to be done.

Regardless of these variables, however, the practitioner has little choice other than to follow the mandates increasingly being established by various regulatory bodies.

The Federal Energy Administration's recommendations, shown in Fig. 12-1, for insulation vary with geographic locations. The Federal Housing Administration's minimum property standards set forth the maximum U values for specific building sections, as shown in Table 12-3.

Many individual states are setting up their own regulations, by incorporating them either within their own building codes or sometimes within some of their administrative agencies, as shown by the minimum insulation standards issued by the New York Public Service Commission listed in Table 12-4.

Proper application of the insulation is as important as the thickness. The National Association of Home Builders (NAHB) points out that proper installation includes making sure that:

- When eave vents are used, there is clearance at the outside plate between the top of the insulation and the underside of the roof sheathing.
- Insulation blankets butt snugly where they meet.
- Separate pieces of blanket are applied and snugly butted at framing offsets.
- Blankets are adequately stapled to eliminate gaps between the insulation and framing members.
- Areas between rough framing and door and window heads, jambs, and sills are filled with insulation.

Table 12-3 FHA maximum U values of ceilings, walls, and openings

		Masonry Wall Construction		Frame Wall Construction		
Winter Degree-Days*	Flat Roof Deck†	Ceilings	Walls	Ceilings	Walls	Doors and Windows
4500 or less	0.14	0.05	0.10	0.08	0.08	1.13
4501 to 8000	0.09	0.05	0.10	0.05	0.08	0.65
4001 or more	0.09	0.04	0.10	0.04	0.08	0.65

Maximum U value of floor sections over unheated basements, unheated garages, or crawl spaces

Winter Degree-Days	Structural Slab	Wood and Metal Framing
2500 or less	†	†
2501 to 4500	0.15	0.10
4501 or more	0.12	0.08

*Winter degree-days are based upon the difference between 65°F and the mean daily outdoor temperature.

†Indicates construction wtih rigid roof insulation and exposed structural system. When roof cavity is available, use column for ceilings.

Source: Minimum Property Standards, The Department of Housing and Urban Development.

R values	Batts or blankets: Glass fiber (in)	Batts or blankets: Rock wool (in)	Loose fill (poured in): Glass fiber (in)	Loose fill (poured in): Rock wool (in)	Loose fill (poured in): Cellulosic fiber (in)
R-11	3½ - 4	3	5	4	3
R-13	4	4½	6	4½	3½
R-19	6 - 6½	5¼	8 - 9	6 - 7	5
R-22	6½	6	10	7 - 8	6
R-26	8	8½	12	9	7 - 7½
R-30	9½ - 10½	9	13 - 14	10 - 11	8
R-33	11	10	15	11 - 12	9
R-38	12 - 13	10½	17 - 18	13 - 14	10 - 11

Heating zone	Recommended for: Ceiling	Recommended for: Floor
0,1	R-26	R-11
2	R-26	R-13
3	R-30	R-19
4	R-33	R-22
5	R-38	R-22

FIG. 12-1 Insulation recommendations. *(Federal Energy Administration, "Tips for Energy Savers.")*

Table 12-4 Public Service Commission maximum U values

	One or Two Family	Multifamily
Roofs	0.05	0.05
Exterior walls	0.07	0.07
Floors over unheated crawl spaces	0.08	—
Floors open to outside or unheated garage	0.05	0.05
Foundation walls	—	0.08
Floors over unheated basement	—	0.08
Glazing	0.69	0.69
Entrance doors	0.40	0.40

Source: N.Y. State Public Service Commission Regulations, 1978.

- Insulation is placed behind electric boxes and on the outside of exterior wall piping.
- Rips and tears in vapor barriers are repaired.
- Other NAHB recommendations include the following:

Mechanical/Electrical Equipment

Clock Thermostat: The installation of a clock thermostat to reduce temperatures at night can result in significant energy savings which vary by type of heating system and geographic location. Temperature setback is usually not recommended in most climates when a heat pump is used.

Water Heater: The water heater is usually the second largest energy user in the house. Locate it as near as possible to the point of greatest use. Investigate the availability of more energy efficient water heaters. Make sure that the temperature setting is as low as practicable (120°F to 125°F). Suggest to the purchaser that it be left at that setting.

Mechanical/Electrical Equipment: Mechanical equipment should be properly sized. Oversizing costs more, uses more energy, is less efficient and produces lower comfort levels. When selecting appliances and mechanical/electrical equipment, consider the comparative energy efficiency of each. The energy efficiency ratio (EER) is a commonly used efficiency rating method for air conditioning. The higher the EER, the greater the energy savings. Until recently, the typical EER for residential equipment was in the 6 to 7 range. Now, equipment having an EER in the 7 to 8 range is common. Some equipment is available with an EER of 9 or slightly higher. Most of these are in the larger sizes.

When using electric heat, consider using the heat pump. The seasonal performance factor (SPF) for a heat pump is greater than electric resistance heat and therefore results in lower operating costs in most areas of the country. In areas of both hot and cold climate extremes, consult the local power company for advice on the merits of heat pump use.

Several newer mechanical/electrical devices to conserve energy are becoming available. These include: automatic stack dampers which close the flue when the furnace is off; intermittent ignition devices for gas furnaces and ranges; heat exchangers to capture waste heat from air conditioners to heat water; and solar water heating or space heating systems in areas where they are justifiable.

Duct System: Avoid running ducts through non-conditioned spaces. If this is not possible, then insulate the ducts. Avoid undersizing duct work. Use of more efficient fittings and minimizing turns increases system efficiency. To reduce duct leakage, keep duct runs short and insure that all joints are properly taped. Tape the joints of metal ducts even if they are to be covered with insulation. As the thermal efficiency of the envelope improves, simpler, lower cost duct systems become possible.

Hot Water

The Plumbing Manufacturer's Institute estimates that a typical family of four uses an average of 255 gal of water each day for interior plumbing. It is broken down as follows:

Dishwashing	15 gal
Cooking/drinking	12 gal
Utility sink	5 gal
Laundry	35 gal
Bathing	80 gal
Bathroom sink	8 gal
Toilet	100 gal

A typical shower takes 5 min and uses between 5 and 6 gal/min of water. When multiplied by a year, the family usage totals 93,000 gal/yr of water; of this, 18,000 gal is hot water. Use of the volume controllers described in Chap. 7 will reduce this quantity by between one-third and one-half.

Lighting, Cooking, and Appliances

As with other types of buildings, the home should be lit no more brightly than necessary, and wherever possible, fluorescents should be substituted for incandescent lamps.

The National Energy Act of 1978 requires the Department of Energy to set efficiency standards within the next two years for 13 categories of home appliances, including refrigerators, dishwashers, air conditioners, television sets, and kitchen ranges.

Estimates for annual energy requirements for existing electric household appliances are shown in Table 12-5.

The Oil-Fired Furnace

A heating system has four principal parts: burner, furnace or boiler, heat distribution system, and chimney.

The burner generates heat by burning fuel oil. Part of the heat produced by the burning of the fuel is absorbed by the furnace or boiler and transferred to

Table 12-5 Annual energy requirements of electric household appliances

	Estimated Kilowatthours Used Annually
Major Appliances	
Air conditioner (room)	860
(Based on 1000 hours of operation per year. This figure will vary widely depending on geographic area and specific size of unit)	
Clothes dryer	993
Diswasher—including energy used to heat water	2100
Dishwasher only	363
Freezer (16 ft_3)	1190
Freezer—frostless (16.5 ft_3)	1820
Range with oven	700
With self-cleaning oven	730
Refrigerator (12 ft_3)	728
Refrigerator—frostless (12 ft_3)	1217
Refrigerator/freezer (12.5 ft_3)	1500
Refrigerator/freezer—frostless (17.5 ft_3)	2250
Washing machine—automatic (including energy used to heat water)	2500
Washing machine only	103
Washing machine—nonautomatic (including energy to heat water)	2497
Washing machine only	76
Water heater	4811
Kitchen Appliances	
Blender	15
Broiler	100
Carving knife	8
Coffee maker	140
Deep fryer	83
Egg cooker	14
Frying pan	186
Hot plate	90
Mixer	13
Oven, microwave (only)	190
Roaster	205
Sandwich grill	33
Toaster	39
Trash compactor	50
Waffle iron	22
Waste disposer	30
Heating and Cooling	
Air cleaner	216
Electric blanket	147
Dehumidifer	377
Fan (attic)	291
Fan (circulating)	43
Fan (rollaway)	138
Fan (window)	170
Heater (portable)	176
Heating pad	10
Humidifier	163
Laundry	
Iron (hand)	144
Health and Beauty	
Germicidal lamp	141
Hair dryer	14
Heat lamp (infrared)	13
Shaver	1.8
Sun lamp	16
Toothbrush	.5
Vibrator	2
Home Entertainment	
Radio	86
Radio/record player	109
Television	
Black and white	
Tube type	350
Solid state	120
Color	
Tube type	660
Solid state	440
Housewares	
Clock	17
Floor polisher	15
Sewing machine	11
Vacuum cleaner	46

Note: When using these figures for projections, such factors as the size of the specific appliance, the geographic area of use, and individual use should be taken into consideration.

Source: "Tips for Energy Savers," Federal Energy Administration, 1977.

air or water, which is then distributed throughout the home by air ducts or hot-water pipes and radiators. The heat that is not absorbed by the furnace or boiler is lost up the chimney in the process of disposing of smoke and gases. The overall heating system efficiency depends on the performance of each of these parts.

Burner Efficiency

Oil must be thoroughly mixed with air to burn completely. This usually requires more air than is actually needed to convert the carbon and hydrogen in the fuel to carbon dioxide and water, which are the products of complete combustion. The amount of this excess air can be determined by measuring the amount of carbon dioxide in the flue.

Generally, the higher the carbon dioxide level, the less excess air used and the more efficient the combustion process. A carbon dioxide level of 9 percent is considered good. Levels over 11 percent are excellent. In general, existing burners should produce efficiencies of 65 percent or better; new burners should show at least 75 percent efficiency.

Furnace or Boiler Oversizing

A technician can determine if a system is oversized (most existing systems are). The technician does this by taking into account the average daily temperatures in the region, the amount of oil used, and alternative nozzle sizes. By switching to a smaller-size nozzle, the system will run longer but burn less oil per unit of time, and the amount of heat lost up the chimney will be reduced.

Furnace or Boiler Efficiency

This check is made by measuring the temperature of the flue gas leaving the furnace or boiler. Flue gas temperature should be from 300 to 400°F for a new boiler and 400 to 500°F for conversion burners. Excessive temperature, measured after the burner nozzle has been properly adjusted, indicates that either the burner nozzle size is too large and more heat is being generated than can be utilized in the heat exchanger, or the heat exchanger surfaces are badly sooted and need brushing and vacuuming. Installation of a fuel oil solenoid valve will help prevent this sooting.

Heat Distribution Efficiency

On air systems, have a technician check and reset, if necessary, the on and off temperature settings of the furnace fan. To conserve fuel, the fan should shut off when the furnace temperature is about 90°F. It should not go on again until the burner comes on and raises the furnace temperature to about 110°F.

Chimney Efficiency

Most chimneys or vents produce more draft than is necessary. It is the job of the draft control to prevent this. Consider installing a barometric draft damper if the system does not already have one.

Future House Design

The house of the future is not likely to be dramatically different from our homes of the past, since it is unlikely that energy considerations will override all others, including function, aesthetics, etc.

Nonetheless, it is of interest to group the anticipated impacts of energy conservation on new housing design. Not all these trends will be incorporated into every home, but increasingly some will show up as part of the overall design concept.

- Housing Type
  Smaller, higher-density, fewer detached houses
  Increased shift to townhouse and low-rise from single-family detached
  Diminished relative attractiveness of mobile homes in life-cycle cost terms
  Improved designs with lower unit demands
- Architectural Features
  Thicker wall (cavity and sandwich) and roof construction for more insulation
  Fewer picture windows, more double- and triple-glazed windows, some specially coated glass
  More functional windows—designed as passive solar collectors
  Tighter, better-sealed joints, higher-performance sealants, better workmanship
  Better control of moisture to protect insulation
  More attention to shape, orientation, landscaping in design
  Control of air movement between floors—fewer open stairwells and split-level designs
  Insulated foundation walls in cold climates
  Greater thermal resistance for more expensive fuels, such as electric heating
  Better thermal comfort and greater acoustical privacy
- Mechanical Systems
  Smaller, more efficient HVAC equipment, better load matching
  Customized ventilation to provide outdoor air when and where needed
  Widespread use of heat pumps in moderate climates—integrated heat pump and solar heating
  Solar space heating in selected climates where gas and oil are in short supply
  More zoning and multipoint control systems in larger residences
  Electric load management control options, for hot water and appliances
  Solar water heating in the South
- Institutional
  Life-cycle cost-based performance standards (voluntary or mandatory) for new housing design
  Labeling of houses, equipment, and appliances for energy use, cost, and performance
  Householders knowledgeable in how to operate homes efficiently

Chapter 13

Case Studies in New Construction

The case studies chosen for this chapter represent a wide range of approaches to the requirements of energy-conscious design. They range from structures which retain their traditional (pre-energy crisis) form and react to the conservation ethic only in the design of mechanical equipment, to those buildings whose form and design reflect the needs of conservation.

What all this will come to, only time will tell. On the one hand, if the energy crunch continues (as now seems likely), its impact on design will be increasingly apparent. On the other hand, if some now unforeseen scientific breakthrough solves the problem, those designs reflecting the energy needs of the building may come to be seen as historical anachronisms, the results of a passing cloud. Only hindsight can be 20/20.

Case 1: GSA's Energy Prototype; Architect—Nicholas Isaak and Andrew C. Isaak of Manchester, New Hampshire. The design of this federal office building has become a classic in the development of energy-conserving structures. Design was started in 1972 and followed traditional procedures between the architects and GSA. At that time, typical GSA buildings had U values of 0.15 for the roof, 0.16 for the walls, and 1.13 for glazing (0.25-in plate glass). The Manchester building was initially designed for −10°F outdoor design conditions in winter, with 75°F interior temperatures. In summer, it was designed for 90°F dry-bulb and 73°F wet-bulb temperatures, with interior requirements of 75°F dry bulb and 5 percent relative humidity. It used venetian blinds for solar control and had lighting levels varying between 3 W/ft² in the toilet and storage areas and 6 W/ft² in office spaces.

The building program called for 126,000 ft² net with office space for six major agency tenants and a variety of smaller ones. Its simulated energy performance indicated that it would need an annual energy input of 116,000 Btu/ft² at the building (representing 216,000 Btu of primary source energy).

Then, suddenly, GSA told the architects to stop work, proceeded to introduce an energy conservation consultant (Dubin-Mindell-Bloome of New York City), and finally set a goal of 55,000 Btu/(ft² · yr) at the building.

The result? Back to the drawing boards. The changes produced by the restudy are shown in Fig. 2-1.

In the final design, the earlier rectangular shape was replaced by a cube to reduce the surface-to-volume ratio of the structure. The first floor is larger to accommodate activities dealing with large numbers of the public, and it measures 75 ft × 200 ft. The second and seventh floors are the smallest, 100 ft × 120 ft, and the overhang provided by floors three through six provides the opportunity for an experiment in daylighting on the second floor by sheltering the greater amount of glazing there. Floors three through six are 118 ft × 138 ft, or nearly a perfect square.

Interior design conditions were changed to call for 68°F in the winter and 78°F in the summer with 60 percent humidity. Temperatures in corridors, toilet rooms (which use activated charcoal filters to halve air exhaust requirements), and storage areas may drop to 65 or rise to 80°F. It has 0.06 ft³/min of outside air per square foot versus 0.25 for a conventional building. Its calculated winter heat loss for the second through seventh floors was 45,000 to 77,000 Btu versus 278,000 to 684,000 Btu for its conventional counterpart.

Windows represent only 12 percent of the east, west, and south facades; the north facade has none at all. In fact, each of the facades is different and reflects the climate (wind and sun) to which it is exposed. The structure is designed to be 80 to 100 percent efficient in prohibiting summer solar loads and 80 to 90 percent efficient in permitting the lower winter sun to penetrate.

As illustrated in Fig. 13-1 the final building has U values of 0.06 for its roof and walls, and its windows, which are double-glazed with built-in blinds between the panes, have a U value of 0.55. The structure is massive, with 12-in-thick masonry backup walls on the building's perimeter. The roof, which is surfaced with crushed white marble to reflect solar heat, is thicker than structural considerations required. The insulation is applied to the outside of the structure, including the overhangs, so that the walls can be used as a heat sink that absorbs interior heat and then reradiates it to the interior to take the edge off peak loads.

Design conditions for lighting levels were much below those of the original. Offices were designed for 2 W/ft² on the average; lobbies for 2.5 W/ft²; toilet and storage areas for 1 W/ft²; and parking for 0.5 W/ft². Each floor of the building has a different combination of lighting elements which serve as a laboratory where different approaches can be tried in the hope of reducing lighting levels even further than those established for the present design.

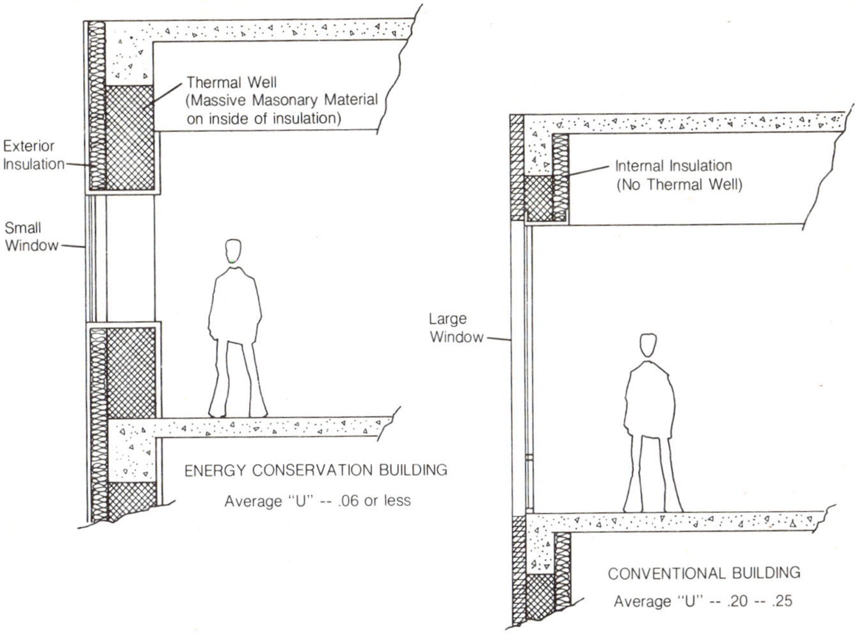

FIG. 13-1 Wall construction of an office building in Manchester, New Hampshire. *(General Services Administration, "A Case Study–Designing an Energy-Efficient Building," 1975.)*

Case 2: 1.6 Million Gallon Energy Bank; Architect—Gordon S. Adamson and Associates. "Hydro Place" is the Toronto headquarters of Ontario Hydro, the government-owned corporation that supplies electricity to Canada's most populous province. The structure is a crescent-shaped, partly glass-enclosed office tower 20 stories in height. The curved-glass facade reflects the surroundings and the passing scene, suggesting a gigantic tapestry suspended in the air. Pedestrians gaze at the ever-changing picture from the landscaped plaza. Shops and restaurants in the underground shopping concourse are visible from the street.

The heart of the energy design for Hydro Place is a huge reservoir in the basement. It was suggested by a Japanese method of storing water in excavations for earthquake protection beneath tall buildings.

Three underground chambers, in a pit 120 ft long by 90 ft wide by 20 ft deep, hold 1.6 million gal of water. Here in one of the largest energy conservation banks in existence as part of a building structure, heat from lights, people, and machines is pumped for storage in the form of hot water. Water kept chilled is used for air conditioning.

Employees in the building, which includes a penthouse for equipment and two levels underground, work with the thermostat set at 73°F without using any fuel for space heating in the Canadian winter. Management estimates that annual energy consumption will be held to 52,500 Btu/ft^2 compared with more than 120,000 Btu/ft^2 in conventional Toronto buildings of similar size and function.

Except for four conference rooms on each floor, interior partitioning has been eliminated. Executives and staff share one enormous room, with portable low screens demarcating work areas, giving everyone much the same outside view through walls of one-way reflecting glass.

The glass used on the front and two sides of the building (the rear is a precast concrete spine from which the glass is suspended two floors off the ground) is set in slender frames of stainless steel with outer and inner panes for insulation. The outer pane is coated on the inside with silver, causing 80 percent of the solar heat to be reflected outward—twice as much as plain glass would deflect—while keeping the interior heat inside.

Interior lighting uses fluorescent units enclosed in reflecting V-shaped recesses in the ceiling, cutting electricity consumption for illumination by one-third. While many of the lights would be kept perpetually lit to provide warmth, the heat storage system in Hydro Place permits the building to be darkened overnight, on weekends, and for even longer periods without a sharp drop in temperature, even in the subzero nights of an average Toronto winter.

A complex of air pumps collects the heat from lights, machines, and human bodies and shifts it around the building. The distribution is managed by a computer with sensors that react when a space is becoming too hot or too cold. Because of the efficiency of the system, the equipment needed for air conditioning, heating, and ventilating is 30 percent smaller than that of equivalent buildings.

Case 3: Underground Well Water; Engineers—Heath Engineering Co. of Salt Lake City, Utah. Water from deep wells both heats and cools the initial three buildings of the Utah Technical College at Orem, Utah, and will also serve future campus buildings. Prior to construction of the campus, a study was made of the relative advantages of various energy services, including coal, oil, natural gas, electric resistance heating, well-water heat recovery, and solar-augmented heat recovery. At the time of the study, there was a good supply of relatively low-cost natural gas where the college was to be built; but the uncertainties concerning its future availability eliminated it as the primary heating medium for the campus.

Calculations indicated that a well-water heat recovery system would require one-third less energy than a comparable fossil-fuel-fired system.

Two identical wells are drilled to a depth of 550 ft. A distance of 800 ft between the wells was as much a function of operating requirements as of campus planning.

The system utilizes water pumped from the deep wells to provide both heating and cooling. In the winter when heating is required, water at 62°F is pumped out of the ground and back to the ground through a heat exchanger. It provides energy to feed the evaporator side of centrifugal chillers used as heat pumps.

In warm weather the same well water is used to cool the condenser side of these units. This is similar to the way household refrigerators produce heat at the back while creating a cold environment inside.

Energy can be recovered from exhaust air or from energy-producing machinery. When internal heat gain is insufficient to satisfy heating needs, the well water is brought into service. This heat source provides energy at a useful temperature through the refrigeration cycle. In the summer, when cooling is required, the waste heat may be rejected to the well water.

In the intermediate seasons of the year when both heating and cooling are needed, the heat recovery system (which includes a heat pump) has equal capacity for either need. Either heating or cooling is dominant, and other energy is available as a no-cost by-product. The system avoids the need for cooling towers and standby sources of energy. Heat generated by people, lights, and the sun can be shifted to compensate for heat loss at the building's perimeter while the system runs minus well water.

The initial annual electrical cost for the college's three buildings was $90,000. The gas cost, including the auxiliary heat furnished by a gas boiler used weekends and nights, is $11,000. These figures represent the servicing of a 192,000-ft^2 campus.

In keeping with the technical teaching aspects of the college, electric transformers were located inside the building, and most of the mechanical equipment and ducts were left exposed. The campus is being built in several phases, funded by the state legislature on a pay-as-you-go basis.

The architect for the campus master plan was Bruce Jenson, Salt Lake City. Frank Ferguson of Salt Lake City designed the Automotive Trades Building and the Student Center; Lee Knell of Provo designed the Business Education Building.

Case 4: Environmental and Energy-Conscious Arboretum; Architect—Malcolm Wells of Cherry Hill, New Jersey. The Cary Arboretum Administration-Research Building is part of the New York Botanical Garden (Fig. 13-2). The building provides botanical research facilities for permanent staff and for research fellows, as well as ecological and botanical educational facilities. The building is two stories high with a gross area of 28,000 ft^2. Construction is masonry and heavy timber.

An energy consumption goal of 50,000 Btu/(ft^2 · yr) was set for the building line, with a goal of 100,000 Btu for raw source energy. Other goals were set for water consumption, hot-water use, air conditioning, and lighting.

In order for the energy goal to be reached, energy-conserving features were introduced including:

- Earth berms
- High mass in walls
- Southern orientation of major openings
- Operable windows with insulating glass
- Insulating shutters on all windows and skylights

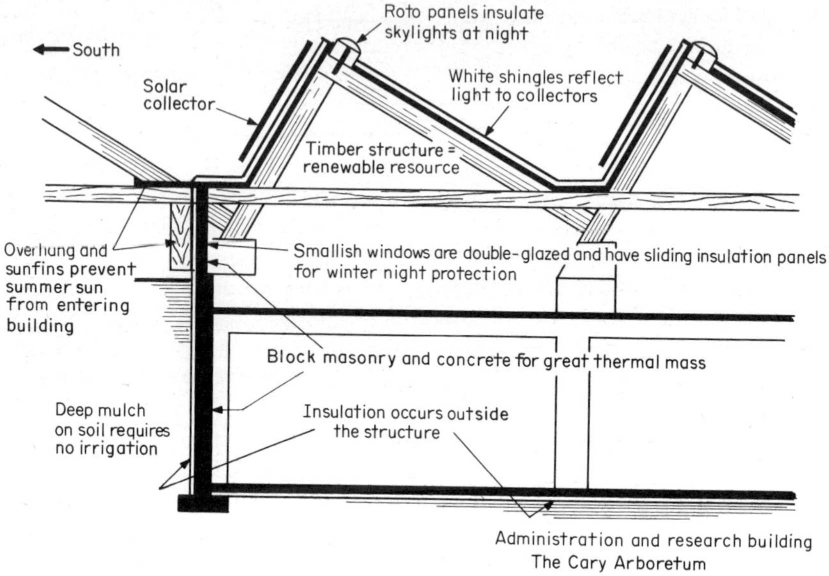

FIG. 13-2 The Cary Arboretum. *(AIA Energy Notebook.)*

- Exterior shading for south windows
- Daylighting with task lighting design, light walls, and surfaces
- Solar space heating
- Solar domestic water heating
- Heat pump
- Cooling with well water
- Automatic control and monitoring system

The chief exterior feature of the building is its saw-toothed roof. This roof form answers several needs. First, it provides the necessary tilted south-facing surfaces for the solar collectors. Second, white shingles are used on the north-facing slope to increase the effectiveness of the collectors; additionally, these same surfaces contain a number of skylights near the ridge line.

Window areas are about 13 percent of the exposed wall; they are maximized on the south, minimized on the north. Daylighting is provided by skylights and windows in many areas during most daylight hours. Task lighting was designed for 2.5 W/ft^2 and makes use of higher-efficiency fixtures such as fluorescent and halide wherever possible.

All windows and skylights are equipped with movable (internal) insulating shutters, which are closed at night during the heating season. South-facing windows are provided with overhangs and side fins to maximize winter sun and minimize summer penetration.

Heavy concrete and masonry walls were chosen to boost the thermal inertia of the structure. The wall construction is insulated on the exterior and covered with cement plaster. The U value is 0.065 above grade, 0.043 below grade.

The building is located on the site with solar exposure and wind protection as major considerations. Major facades are oriented either north or south. The building is positioned against heavy woods to the north and west for protection against winter winds.

About two-thirds of the exterior walls are in contact with the earth. Most of the lower level (library, laboratories, mechanical spaces) is below grade. Earth berms cover many wall areas above the upper floor line, where offices, display space, meeting rooms, and laboratories are located.

Case 5: Form Follows Energy; Architect—Richard L. Crowther of Cherry Creek, Colorado. Located on a corner site on the fringe of a retail area, this project contains two almost identical buildings, each two stories high and containing 4500 ft^2 each (*Fig. 13-3*). It illustrates how the designer can take a particular part of the program (in this case, its energy needs) and have it make a significant contribution to the architectural design.

The combination of collector, clerestory window, and reflector serves many purposes. In winter, the clerestory window admits direct sunlight as well as (diffuse) light reflected from the white mica stucco surface of the window overhangs. During the summer when the sun is high, the overhang shades the clerestory window from direct sun. Light enters only after reflection from the overhang.

The roof itself is surfaced with white marble chips, increasing winter heat gain and decreasing summer heat gain. In winter when the sun is low, the marble chips reflect diffuse radiation into the collector, increasing the amount of heat at the absorber plate. Light is also reflected into the clerestory window after a double reflection from the roof and then the overhang. These diffuse reflections increase the effective area of the solar collector by as much as 15 to 20 percent. In the summer, when the sun angle is high, a large portion of the

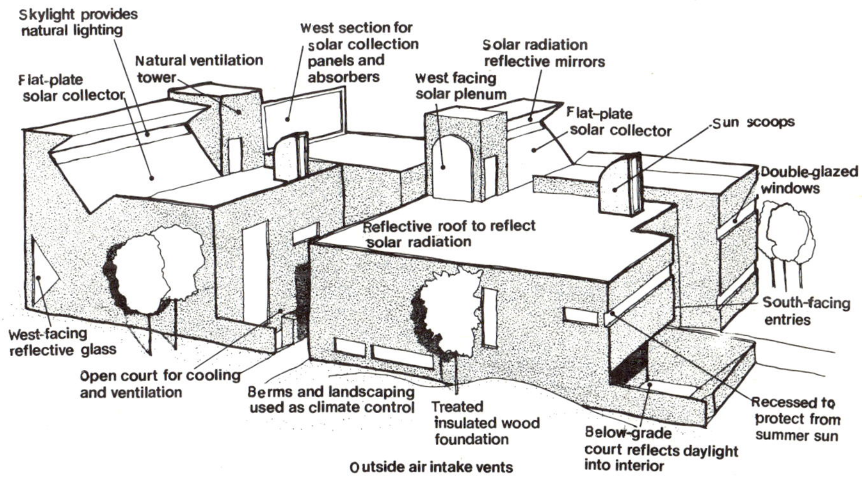

FIG. 13-3 Energy as design determinant. *(AIA Energy Notebook.)*

diffuse radiation is reflected to the sky. Exterior walls are darker than the roof to balance year-round gains and losses.

During planning of the project, the architect's office had an interesting exchange with the local utility company. The utility originally proposed to levy a 50 percent penalty because it said that as a solar structure the building would use the utility's energy only intermittently and therefore should be billed at standby customer rates. In a letter to the utility the architects explained that the buildings would not be standby customers, but rather would have an almost constant daily and year-round demand for electricity, with no peaks for auxiliary heating. The letter included a list of all the electric equipment and explained the system of interlock switches that would limit peak demand:

> We have designed this building to put as constant and low a demand as possible on the Public Service Company. . . . This building will be using 80 percent less energy than a conventional building, but will have a utility bill which is only 10 to 15 percent less. This 10 to 15 percent savings is not sufficient to prompt many builders to incorporate energy conservation into their designs. We encourage the Public Service Company to enjoy a fair profit; yet we also think that they have a socio-economic responsibility to distribute the cost of benefits to all their customers, when all their customers are benefited, such as by energy conserving and solar buildings . . . through energy conservation and wise energy management we are actually a model Public Service customer, the type you should encourage. Our year-round demand is constant, our peak demand is low, and our "auxiliary" equipment demands are met by switching off interruptible equipment.

The upshot of this correspondence with the utility was that the 50 percent penalty was waived.

The exchange is typical of what may be expected whenever pioneering efforts run up against a bureaucracy, be it in government or a public utility.

Case 6: Refuse-Fired Power Plant; Architect/Engineer—Glaus, Pyle, DeHaven Associates of Akron, Ohio. Hooker Chemical Corp. will be the first U.S. manufacturer to generate steam and electricity from a mixed bag of municipal and commercial garbage when its Niagara Falls, New York, energy center goes into operation, currently scheduled for 1980.

Should the supply of garbage be interrupted for any reason, the plant is designed to burn either coal or oil instead. The Hooker installation is applicable to certain major manufacturing plants located near large population centers and offers benefits to both the manufacturer and the local community. These benefits include:

1 Freeing valuable fossil fuels for other uses in the area

2 Providing a local and dependable source of fuel for the plant at reasonably predictable cost

3 Stabilizing refuse disposal costs for the community

At Niagara, for example, energy released from refuse will provide all the steam and a significant portion of the electricity the company requires, decreasing the company's fuel oil consumption by more than 15 million gal/yr. Hooker has entered into contracts with a landfill operator who will deliver 2200 tons/day of refuse for the facility's two boilers. Hooker may install a third boiler in the future if additional refuse can be obtained under contract.

Garbage is scheduled for delivery to the plant by truck 8 to 10 h/day, 5½ days/wk. After trucks are weighed and tipping fees are collected, the garbage is dumped from one of 14 unloading stations into a 1600-ton storage pit. Two hydraulic multiram unloaders installed at the bottom of the pit transfer up to 160 tons/h to the surge hoppers serving the shredder-feed conveyors. Either unloader is capable of supplying the 2200 tons/day required by the two-boiler plant.

From the shredders the garbage is carried to the air classifiers. The air classifiers separate the light, combustible material from the heavier, shredded refuse stream, thus producing fuel for the boilers while permitting recovery of potentially valuable dense materials such as metal and glass.

Two belt conveyors, operating in parallel, transfer the refuse-derived fuel from the air classifiers to the storage bin, from which it is conveyed to the boilers.

The steam generators are almost duplicates of the boilers used by pulp and paper companies to burn bark. They are designed to produce 300,000 lb/h of 1250-lb/in^2 gauge steam at 750°F on a continuous basis. Both boilers also can generate 300,000 lb/h of steam at rated temperature and pressure when burning either coal or no. 6 oil. Hooker, however, does not anticipate burning either coal or oil unless its refuse supply is interrupted for more than two days or problems with the refuse preparation and handling system demand use of an alternate fuel.

Case 7: Office Building with Storage Balance Wheel. The thermal storage potential of a reinforced concrete structure and a large rock bed, both cooled with night air, is used in conjunction with a tight building skin, operable exterior sunshades, and greatly reduced artificial lighting to produce an energy-efficient office building for the state of California (in Sacramento). (See Fig. 13-4.) This building is projected to require less than 20 percent of the energy consumed annually by existing state and office buildings. The rock bed is used for heat recovery and can be operated to reduce the demand on the mechanical chilling equipment by 75 percent. It is estimated that the night ventilation of the structure will meet 65 percent of the total cooling while the rock bed will provide an additional 25 percent. Total gross energy consumption for the building is estimated to be 64,300 Btu/(ft^2 · yr); the building has an area of 267,000 ft^2.

The building is composed of two L-shaped sections wrapped around a central covered atrium. This passively conditioned space houses the main vertical and horizontal public circulation, the two thermal storage rock beds, and public sitting areas; and it serves as a preconditioner for the building's makeup air. The majority of the building's reinforced-concrete structure is directly exposed to the thermal environment of the occupied office space and is cooled at night by evaporatively cooled night air circulated through the central variable-air-

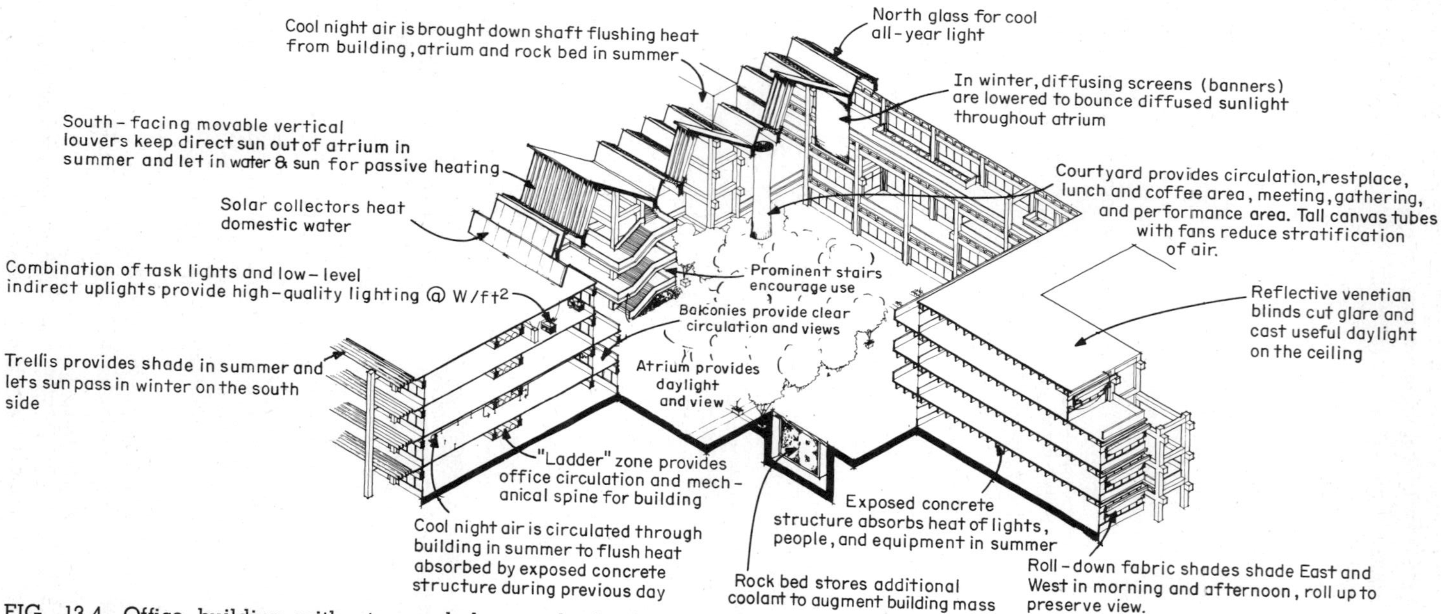

FIG. 13-4 Office building with storage balance wheel. [*Proceedings of the 2d National Passive Solar Conference, March 16–18, 1978; the American Section of the International Solar Energy Society (ISES), vol. 2.1.*]

volume (VAV) system. The rock beds, also purged with evaporatively cooled night air, are sized to meet 60 percent of the daytime cooling load on the cooling design day and will provide an estimated 75 percent of the cooling not otherwise met by the thermal storage of the building's structure.

The major energy design objectives of the project were to reduce the peak heating and cooling loads, to shift as much of the energy consumption as possible to off-peak hours, to reduce the total energy use, and to provide effective integration of the energy systems with the other architectural and user-related requirements of the building. In evaluating the dynamic behavior and cost-effectiveness of various possible solutions, a range of problems were considered:

- The costs of various building elements (value of rentable space, cleaning and maintenance costs, the need for flexible office size and layout, HVAC costs)
- The dynamics of large buildings (discontinuous occupancy and environmental control requirements, large intermittent internal lighting and equipment loads, multiple zones, and relatively tight allowable temperature ranges)
- The relationship between the users and the building (degree of user participation acceptable, acoustical treatment of open-plan offices, extent of individual comfort control, windows as amenities rather than energy elements)

In many cases, these considerations combined to eliminate options that were, in energy terms alone, very workable. Final energy decisions included the following:

Lighting

Artificial lighting loads had a maximum of 1.0 W/ft^2 for perimeter zones and 2.5 W/ft^2 for interior zones. The final design uses ceiling-mounted up/down fluorescent lights in enclosed offices and high intensity discharge (HID) up lights with individual task lights in open-plan areas.

Building Envelope

Elements included double glazing, R-20 walls and roof, as well as the use of the atrium as an entrance vestibule. Sun control is provided by operable exterior roller shades on the east and west facades and a fixed louvered overhang on the south. North-facade shading was found not to be cost-effective.

Cooling System

The cooling design is based on the large diurnal temperature swing characteristic of Sacramento's cooling season. The basic strategy involves the use of thermal mass elements to absorb and store heat generated during the day and the use of cool night air to dissipate that heat. Thermal mass is provided by two main elements—the structural mass and rock-bed storage. The building's structural mass is equivalent to 1 ft^3 of concrete with a 3.3-ft^2 surface for every square foot of occupied floor space. Rock-bed storage is provided by two separate beds, each of which is connected via vertical air shafts with the stack of mechanical rooms adjacent to it. Each bed holds approximately 1.35 million lb of 1-in (average) rock dredged from a dry channel of the nearby American River. The frontal area of each bed is 2000 ft^2, and the depth is slightly under 7 ft.

Case 8: Computerized Mechanical Systems; Architects—Thompson, Ventulett, Stainback & Associates of Atlanta, Georgia. This structure is the headquarters for the Atlanta division of IBM. As such, it makes complete use of computerized controls. The building was under construction during the 1974–1977 period. The energy management system is controlled by an IBM System 7 computer.

The computer automatically turns off lights at specified times, regulates air conditioning and heat, and monitors and controls humidity and temperature by zone. It uses data collected by the building's own rooftop weather station to control interior temperature and humidity. The System 7 monitors more than 300 sensors attached to chillers, boilers, air handling systems, and cooling towers to ensure that energy is not wasted.

The 100 ft × 400 ft structure spans two hillsides and is suspended above an artificial lake that serves as a reservoir to control rainwater runoff. The building stands 11 stories above the lake.

The building was dug into the hillsides, so that the lower floors are well insulated by the earth and less of the building is exposed to the elements. Total glazing of the architectural building is 38 percent. In order to get more control of the solar heat, large areas of floor-to-ceiling glass were shaded by 10 ft of overhang—poured-in-place concrete spandrels. The overhang is supplemented by redwood sun screens suspended under the spandrels. On the dining terrace, the glass is set well back, from 7 to 15 ft in the deep recesses.

The folded, sawtooth configuration of the glass prevents some areas from getting any afternoon sun. Of the glazing, 80 percent is insulated glass which absorbs some solar heat, but also does not obscure the view.

During the winter months, the building's electric boiler is seldom used, since heat generated by the computers in the facility is used to heat the structure. The heat recovery system utilizes a heat recovery chiller. Only when the temperature gets below 11°F is it necessary to start the boiler.

As originally planned, the lighting for the facility was based on four-tube fluorescents. During construction these were changed to three-tube fluorescents with a corresponding energy saving and no loss of efficiency in actual use.

Case 9: Underground School; Architect—Davis, Smith & Carter of Reston, Virginia. This school building was designed to be set underground, and it takes its name, Terraset, from this design (Fig. 13-5). Construction of the 69,000-ft^2 building began with the removal of the top of a low hill. The building's concrete

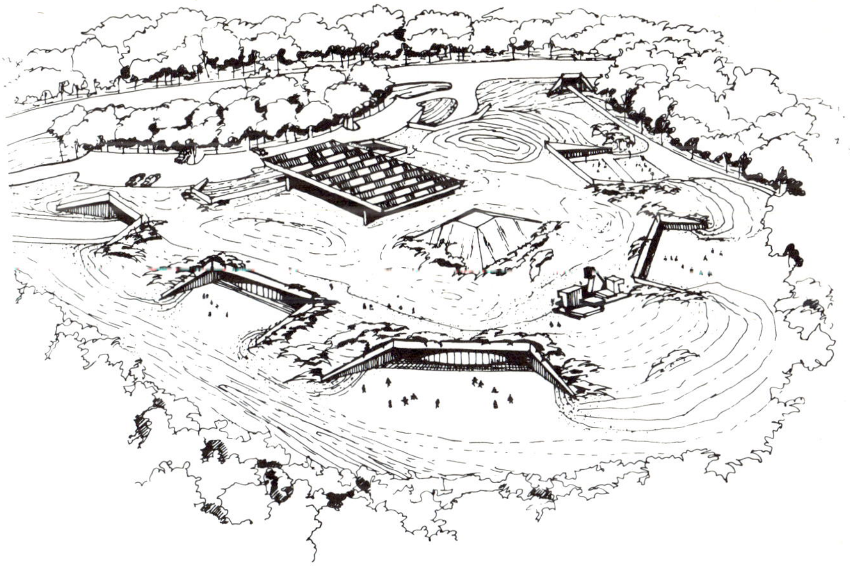

FIG. 13-5 Underground school. *(AIA Energy Notebook.)*

shell was then poured and covered with earth on the top and sides to a depth of 3 to 4 ft.

As constructed, the building has three main features that contribute to reduced energy consumption:

- *The earth cover.* This cover serves as natural insulation. The earthen mass forms a reservoir for warmth or coolness, delaying the effect of outside temperature changes. It acts as a thermal flywheel.
- *Heat recovery.* Heat from people, lights, and equipment is reclaimed and used for heating purposes. The heat recovery system is expected to reduce the heating bill by 50 percent.
- *The solar heating and cooling system.* This system employs evacuated tube collectors to supply water at up to 240°F to two absorption chillers, piped in series with an electrically driven reciprocating chiller for cooling the building. The solar-heated water is also used directly for heating.

One enters the school through a below-ground courtyard sheltered by solar collectors with offices on one side and a multipurpose hall on the other. The bulk of the school proper consists of four circular learning centers (classrooms) around a media center (library), which has a skylight. Since there are windows where the earth is not bermed up, the outdoors can be seen from each learning center.

Several precautionary steps were taken to guard against a public relations backlash on the underground issue. First, the project was given its Terraset name to offset any mole image. Second, elaborate measures were taken to prevent any leaks. This included damming and flooding the roof again and again, to test watertightness. Third, because 20 percent of the school wall area is glass, it is always possible to look outside. Contributing to the ability to look outside are elevated inside walkways and glass-topped class partitions.

In addition to the three main energy-conserving features listed, the design also provided these features:

- Windows were recessed to keep out solar gain.
- The circular floor plan reduced the amount of exterior skin.
- The energy consumed to satisfy the cooling load also heated the perimeter, provided domestic hot water, and raised the temperature of storage tanks for provision of nighttime heating.

Chapter 14

Case Studies in Retrofitting

The case studies included in this chapter were selected to represent the wide range of approaches to energy conservation that arise when existing construction is examined. In almost every instance, the work described has a payback of a year or less and thus stands an excellent chance of being implemented. Work with longer paybacks requires an investment, and the capital needed is usually hard to find. In this respect, retrofit work is different from conservation strategies for new buildings, since the latter can usually cover the cost of the conservation work within the overall financing strategy being used for the new construction. In new construction, conservation is but one of a host of items financed; in retrofit work, conservation stands alone and must make it on its own.

Case 1: Federal Office Buildings. In response to the oil embargo of 1973–1974, the Federal Energy Administration initiated a lighting and thermal operations management program which could be instituted by in-house operating and maintenance personnel. Essentially the program involved the resetting of winter and summer thermostats, together with a cutback in lighting levels. Energy savings averaged 27 percent. The greatest cost was for lamp removal, and this one-time, initial cost amounted to about 20 percent of the first-year savings. (See tables 14-1 through 14-4.)

Case 2: Insurance Office Building. The New England Life office building in Boston occupies 660,000 ft^2 in 10 stories. Purchased steam is used for heating and air conditioning. Electric load goes to lighting, fans, motors, and computers, with the computers using a large fraction of the load.

Tapes of electric load were available from Boston Edison, and these were studied to see if peak demand could be lowered. Air-conditioning fans were found to be starting during the top kitchen loads in the morning, for one thing. Orders went out for a start-up schedule for these fans, to reduce peak load. The computers were another problem in regard to peak load; some types could be powered down at times, while others could not.

Building lighting was a good target for conservation. The interior garage lighting was the first subject of experiment. In a small area, light was cut to a level that later allowed installed wattage to be reduced from 24,000 to 9240 without sacrifice of safety or effectiveness. In the rest of the building, light load is only about half of what it formerly was. The power saved is $13.94 per year for

Building 1 (Table 14-1)

Location	: St. Louis, Missouri
Building type	: Office
Area	: 465,775 ft²
Height	: six floors
Built	: 1961

Table 14-1 Building 1: St. Louis, Missouri

		Energy Conservation Program	
		Before (1973)	After (1974)
Illumination			
At work stations	footcandles	60	50
In work areas	footcandles	60	30
In nonwork areas	footcandles	20	10
Lamps	number	24,900	21,050
Thermostat Setting			
Summer	°F	74	78
Winter	°F	74	68
Building Occupancy			
Working	hours	9	9
Custodial	hours	+8	+3.5
Fan Operation			
Weekday	hours	13	12
Weekend	hours	0	0

Savings

kWh consumed in 1973	10,795,000
kWh consumed in 1974	8,284,000
kWh saved	2,511,000 or 23 percent

2,511,000 kWh saved × 2.1¢/kWh = $52,700 for 1974

Source: "Lighting and Thermal Operations: Energy Management Action Program for Commercial-Public-Industrial Buildings," Federal Energy Administration, 1975

Building 2 (Table 14-2)

Location	: Columbia, South Carolina
Building type	: Office
Area	: 97,253 ft^2
Height	: Six floors
Built	: 1952

Table 14-2 Building 2: Columbia, South Carolina

		Energy Conservation Program	
		Before (1973)	After (1974)
Illumination			
At work stations	footcandles	100	50
In work areas	do	100	30
In nonwork areas	do	30	10
Lamps	number	3,260	2,520
Thermostat Setting			
Summer	°F	74	78
Winter	do	74	68
Building Occupancy			
Working	hours	9	9
Custodial	do	+8	+4
Fan Operation			
Week-day	do	12	9
Weekend	do	0	0

Savings	
kWh consumed in 1973	1,077,400
kWh consumed in 1974	911,900
kWh saved	165,500

165,500 kWh saved × 2.2¢/kWh = $3600 for 1974

Source: "Lighting and Thermal Operations: Energy Management Action Program for Commercial-Public-Industrial Buildings," Federal Energy Administration, 1975.

each fixture removed, plus $1.44 for each fixture relamped with new, lower-watt tubes.

Terminal reheat, originally in the system, has been removed and replaced with a variable-volume system. Use of an infrared (thermographic) camera to examine steam traps and other maintenance-sensitive elements of the steam system also contributed to a reduction in steam use from 47,954 to 30,746 Mlb.

Case 3: Two-Story, Wood-Frame Army Office Building.* Building 1000 is a 30-year-old office building located at Oak Ridge National Laboratory, Oak Ridge, Tennessee. It is a two-story, wood-frame structure which was recovered with aluminum siding. Most of the building's 48,000-ft^2 floor area is used as offices for

*This section was adapted from "An Evaluation of Energy Conservation Alternatives in an Engineering Office Building," reprinted by permission of ASHRAE from *ASHRAE Transactions* 1976, vol. 82, pt. 2.

Building 3 (Table 14-3)

Location	: Washington, District of Columbia
Building type	: Office
Area	: 1,844,000 ft²
Height	: Seven floors
Built	: 1951

Table 14-3 Building 3: Washington, District of Columbia

		Energy Conservation Program	
		Before (1973)	After (1974)
Illumination			
At work stations	footcandles	60	50
In work areas	do	60	30
In nonwork areas	do	20	10
Lamps	number	67,580	33,080
Thermostat Setting			
Summer	°F	74	78
Winter	do	74	68
Building Occupancy			
Working	hours	11	11
Custodial	do	+8	+4.5
Fan Operation			
Weekday	do	12	12
Weekend	do	0	0

Savings

kWh consumed in 1973	27,029,000
kWh consumed in 1974	21,336,000
kWh saved	5,693,000 or 21 percent

5,693,000 kWh saved × 2.2¢/kWh = $125,000 for 1974

"Lighting and Thermal Operations: Energy Management Action Program for Commercial-Public-Industrial Buildings," Federal Energy Administration, 1975.

about 200 engineering personnel. Other characteristics of the building which were considered in evaluating energy conservation alternatives are as follows:

1 *Walls and floors:* Wood, with no insulation

2 *Roof:* Flat, built-up, with 3 in of blown wool insulation in the ceiling

3 *Windows:* Double-hung, with aluminum frames

4 *Lights:* 2465 fluorescent bulbs, 40 W each

5 *Ventilation:* 4700-ft³/min exhaust for toilets and blueprint machines

6 *Occupancy:* 8:00 a.m. to 4:30 p.m., weekdays

7 *Heating system:* Electric baseboard heaters

8 *Cooling system:* 160 package air-conditioning units

9 *Equipment:* Miscellaneous small equipment, such as copiers, vending machines, water coolers, and blueprint machines

To determine the effect of various conservation alternatives, yearly energy requirements were established for a base case and nine alternative modes of building operation. Base-case operating conditions were those which existed prior to the energy conservation era. The alternatives considered were:

1 Modified lighting schedule
2 Reduced lighting levels
3 Modified ventilation schedule
4 Equipment shutdown
5 Additional floor insulation
6 Storm windows
7 Reset inside temperature

Building 4 (Table 14-4)

Location	: Harrisburg, Pennsylvania
Building type	: Office
Area	: 204,300 ft^2
Height	: 13 floors
Built	: 1967

Table 14-4 Building 4: Harrisburg, Pennsylvania

		Energy Conservation Program	
		Before (1973)	After (1974)
Illumination			
At work stations	footcandles	100	50
In work areas	do	100	30
In nonwork areas	do	30	10
Lamps	number	15,780	8,280
Thermostat Setting			
Summer	°F	74	78
Winter	do	74	68
Building Occupancy			
Working	hours	9	9
Custodial	do	+8	+3
Fan Operation			
Weekday	do	18	12
Weekend	do	0	0

Savings

kWh consumed in 1973	4,180,000
kWh consumed in 1974	3,010,000
kWh saved	1,170,000

1,170,000 kWh saved × 2.8¢/kWh = $32,700 for 1974

Source: "Lighting and Thermal Operations: Energy Management Action Program for Commercial-Public-Industrial Buildings," Federal Energy Administration, 1975.

8 Daylight savings time—April to October
9 Daylight savings time—Year-round

Calculations were made with a computer which was programmed to consider

1 Average hourly weather data
2 Thermal properties of walls, roof, and windows
3 Area and orientation of structure surfaces
4 Hourly occupancy, lighting, and equipment profiles
5 Internal equipment and lighting loads
6 Thermostat settings
7 Type of heating and cooling systems

Evaluations and Results

Base Case

The building was assumed to be operating as follows: 5 days per week with 8:00 a.m. to 4:30 p.m. occupancy, full lighting on a continuous basis, and heating and air conditioning as required to maintain 75°F year-round. On this basis the annual energy consumption was found to be 2,413,600 kWh. At the $15.16 per 1000 kWh cost of electricity in Oak Ridge, this represented a yearly cost of $36,-500. The distribution of energy among lighting, heating, and cooling is shown in Table 14-5.

Table 14-5 Energy consumption in building 1000 (1000 kWh)

Alternatives	Lighting		Heating		Cooling		Total Electical		
		% Δ		% Δ		% Δ		% Δ	Σ % Δ
Base case	1215.7	—	275.9	—	921.9	—	2413.6	—	—
Case 1	720.9	−40.7	553.5	+100.6	585.8	−36.5	1860.2	−22.9	−22.9
Case 2	526.4	−16.0	628.2	+27.1	434.3	−16.4	1588.9	−11.2	−34.1
Case 3	526.4	—	387.0	−87.4	414.9	−2.1	1328.3	−10.8	−44.9
Case 4	526.4	—	105.4	−102.0	210.6	−22.1	842.4	−20.1	−65.0
Case 5	526.4	—	81.6	−8.6	218.7	+0.9	826.7	−0.7	−65.7
Case 6	526.4	—	63.1	−6.7	221.8	+0.3	811.2	−0.6	−66.4
Case 7	526.4	—	43.7	−7.0	215.9	−0.6	785.9	−1.0	−67.4
Case 8	526.4	—	47.9	+2.0	204.4	−1.2	778.7	−0.3	−67.7
Case 9	526.4	—	57.7	+3.5	199.5	−0.5	783.5	+0.2	−67.5

Case 1: Lighting schedule change
Case 2: Reduce lighting levels
Case 3: Night shutdown of exhaust air
Case 4: Night shutdown of A/C units
Case 5: Add first-floor carpet
Case 6: Add storm windows
Case 7: Reset temperature 78°F/68°F
Case 8: DST, April–October
Case 9: DST, year-round

Source: Oak Ridge National Laboratory, Office of Technology Utilization/Commercialization. Reprinted by permission of ASHRAE from *ASHRAE Transactions* 1976, vol. 82, pt. 2.

Table 14-6 Energy consumption in building 1000 (1000 k Wh)

	Total Electrical			Energy Costs* ($/yr)	Energy Saved ($/yr)
		% Δ	Σ % Δ		
Pre-energy conservation era mode	2413.6	—	—	36,600	—
Modified lighting schedule	1860.2	−22.9	−22.9	28,200	8400
Reduced lighting levels	1588.9	−11.2	−34.1	24,100	4100
Reduced ventilation	1328.3	−10.8	−44.9	20,100	4000
Equipment shutdown	842.4	−20.1	−65.0	12,800	7300
Add first-floor carpet	826.7	−0.7	−65.7	12,500	300
Add storm windows	811.2	−0.6	−66.4	12,300	200
Reset temperature, 78°F/68°F	785.9	−1.0	−67.4	11,900	400
Daylight Saving Time—April to October	778.8	−0.3	−67.7	11,800	100
Daylight Saving Time—Year-round	783.5	+0.2	−67.5	11,900	+(100)

*Electricity costs $15.16 per 1000 kWh (8/1/75).

Source: Oak Ridge National Laboratory, Office of Technology Utilization/Commercialization. Reprinted by permission of ASHRAE from *ASHRAE Transactions* 1976, vol. 82, pt. 2.

Operating Experience

Occupants have experienced slight discomfort as a result of the temperature settings. No adverse effects have been noted as a result of less lighting and ventilation. With equipment shutdown overnight during the summer, offices are a bit stuffy at 8:00 a.m. but not uncomfortable. During the winter, overnight temperatures decrease but are reestablished to a comfortable level once heaters are turned on. After a weekend shutdown during extreme weather, several hours are required to reestablish comfort conditions, especially at floor level. This probably accounts for heaters being left on during the winter months.

The costs of the required modifications have been minimal. Since implementation is mostly dependent on administrative control, the cost involved has been limited to the removal of light bulbs and the installation of timers on the exhaust fans. The total cost of these items is estimated to be less than $2000.

Summary and Conclusions

A summary of all energy conservation alternatives considered along with the resulting savings or costs is shown in Table 14-6. Conclusions are made as follows:

1 Almost 45 percent of the energy formerly used in Building 1000, which is worth $16,500 per year, is being saved without any adverse effects and without excessive costs for modifications.

2 Another 20 percent, worth $7300 per year, is being saved with only slight adverse effects and without excessive modifications.

3 Additional insulation does not seem to be an attractive economic alternative but could possibly be justified on the basis of improved comfort conditions.

4 Resetting thermostats to 68°F in winter is viewed as inappropriate when the resulting discomfort is weighed against the small amount of energy saved.

5 Daylight savings time does not seem to be worthwhile for Building 1000 in east Tennessee.

Case 4: Industrial Plant Conservation. Typical of the conservation efforts of industrial firms is the program put into practice by the Energy Management Section of General Motors Corporation (GMC). By January 1977, more than 1200 conservation case histories had been implemented. They represented annual savings of $3.4 million, achieved at an implementation cost totaling $1.2 million. While the GMC program included many efforts devoted to saving energy in the industrial processes themselves, the following sampling lists activities relating to energy used by building structures.

Item 1: Heat Reclaimed from Roof Monitor Area

Situation Exhaust heat from a casting operation was discharged through a roof monitor by using a 48-in-diameter exhaust fan.

Action Ductwork was installed to divert the exhaust air to an adjacent building where the waste heat could be utilized for building heat. A seasonal damper makes it possible to discharge the heat outside during the summer months.

Result Energy savings amounted to 6687 MMBtu/yr. Cost of the work was $2600; annual savings were over $6000.

Item 2: Electrical Power Demand Monitoring

Situation The demand portion of the electrical bill for one plant amounted to about one-half of the total monthly cost of electricity for the plant.

Action A minicomputer and controls were installed to shut off predetermined equipment automatically during peak demand periods. The computer also was programmed to turn off this equipment when it was not needed on the third shift and on weekends.

Result Energy savings amounted to 7672 MMBtu/yr. Cost of the work was $175,000; annual savings were over $93,000.

Item 3: Computerized Control of Lights

Situation Interior lighting in a remote tire storage area was left on at all times.

Action *The lighting system was connected to an existing computerized facility monitoring and control system. Now the lights are automatically turned out during off shifts, weekends, and holidays.*

Result *Energy savings amounted to 1290 MMBtu/yr. The cost of the work was $2840; annual savings were over $11,000.*

Item 4: Burning Waste Oil in the Powerhouse

Situation *Waste oils and lubricants used to be collected and sold to salvage dealers for a nominal sum.*

Action *The oils were collected and burned in the plant's powerhouse to generate steam. This replaces the equivalent of 93,000 gal of fuel oil annually.*

Result *Energy savings amounted to 13,020 MMBtu/yr. The cost of the work was $1500 (trucking cost); annual savings were over $32,000.*

Item 5: Shutdown of Boilers on Weekends

Situation *Shutting down plant boilers on weekends was believed to be poor operating practice at many facilities.*

Action *At one plant, experimentation showed that many boilers could be shut down when outside temperatures were above freezing. Over 30 additional GMC plants are able to institute similar boiler shutdown programs.*

Result *Savings in steam at this pilot plant amounted to 58,860 MMBtu/yr. The cost of the work was zero; annual savings were over $137,000.*

Item 6: Fan Shutdown—Heating and Ventilating System

Situation *Roof-mounted fans that were part of a plant's heating and ventilating system were allowed to run most of the time because of the difficulty in getting to them.*

Action *Timers were installed on 16 fan motors to turn them off automatically at predetermined times.*

Result *Electrical savings amounted to 925 MMBtu/yr. The cost of the work was $2400; annual savings were over $6000.*

Item 7: Automatic Control of Rooftop Air-Conditioning Units

Situation Eleven rooftop air conditioners serving a large office building were controlled by room thermostats and were often left running during downtime periods.

Action Seven-day timers were installed on each unit and programmed to shut the units down during lunch periods, off shifts, and weekends.

Result A savings in electricity amounted to 2178 MMBtu/yr. The cost of the work was $1100; annual savings were over $15,000.

Item 8: Reducing Speed of Office Building Ventilation Fans

Situation Office workers complained when the temperature in their building was reduced to 68°F.

Action The pulleys on the main air supply fans were changed to decrease fan speed by 20 percent. Workers no longer complained about cold drafts.

Result A savings in electricity amounted to 885 MMBtu/yr. The cost of the work was $300; annual savings were over $5000.

Item 9: Elimination of Heat and Hot Water in an Equipment Storage Building

Situation A building used for storing production equipment was heated and had hot-water facilities even though no personnel were assigned to the building.

Action The heat and hot water were provided by six gas-fired space heaters and one gas-fired hot-water heater. All gas-burning equipment was disconnected and the water was shut off and drained from the lines.

Result A savings in natural gas amounted to 1826 MMBtu/yr. The cost of the work was $500; annual savings were over $3000.

Item 10: Install Self-Closing Faucets on Circular Wash Basins

Situation Sprays on circular wash basins operated on a 360° pattern.

Action One plant eliminated the 360° pattern by installation of four self-clos-

ing faucets on each circular basin. Now only one-fourth of the heated water is used by a single person.

Result *Energy savings amounted to 2646 MMBtu/yr. The cost of the work was $11,000; annual savings were over $6000.*

Item 11: Heat Reduction at a Remote Warehouse

Situation *Heat was maintained throughout a remote warehouse only because it contained a wet sprinkler system and a small inspection operation requiring a few workers.*

Action *The wet sprinkler was converted to a dry system, and local infrared heating was installed at the inspection site. All other building heat was turned off.*

Result *A savings in natural gas amounted to 6669 MMBtu/yr. The cost of the work was $9750; annual savings were over $10,000.*

Item 12: Installation of Solar Film on Office Building Windows

Situation *Heat buildup in an office building with a broad expanse of windows on the west side caused a heavy load on the building air-conditioning system.*

Action *Solar film was installed on the windows to reflect the sun's rays and reduce the heat buildup.*

Result *Electric energy savings amounted to 300 MMBtu/yr. The cost of the work was $5000; annual savings were over $2000.*

Case 5: Window Replacement. The experience of the public school at Bigfork, Minnesota, indicates the benefits to be derived by replacing inefficient windows with double-glazed, thermal-barrier designs. Bigfork is 200 miles north of Minneapolis and can routinely experience sustained periods of −30°F and −60°F wind-chill temperatures.

Bigfork School was built in several segments; the first in 1926–1927; then in 1938 and 1950; and the latest addition in 1970–1971. No matter how high the thermostats were set, it was still drafty through the old windows. Students seated near the windows were cold and had to wear jackets, and it was common to send the children home when temperatures dropped below −30°F. Teachers' desks were moved as far as possible from the windows.

In the summer of 1970, the old windows were replaced with double-glazed, thermal-barrier windows which produced a dramatic reduction in heat loss. In addition to the glazing, the windows have an insulated porcelain-enamel upper panel which reduces the glass area within the old masonry openings. In addition, an integral exterior panning completely covers the old wood to help seal cracks and crevices.

By comparing the heating costs on a square-footage basis, the average cost for the two years preceding installation of the replacement windows was 22.8¢/ft². The average cost for the three years after installation was 14.9¢/ft².

Case 6: Upgrading of Central Boiler Plant. During the 1973–1974 heating season, the two boiler plants that serve the Booker T. Washington Estates housing complex in Jersey City consumed 417,000 gal of heavy fuel oil, which at spring 1975 prices would cost $146,000. Recognizing that a small percentage savings in fuel use could have a significant monetary payback, the Jersey City Housing Authority invited Exxon Research and Engineering Company to make a visual inspection of the heating plants and to recommend steps that might be taken in order to conserve fuel.

The Housing Authority set a requirement that any expenditures must be covered by a cost savings generated within the first year after the work was done.

Given this limitation, Exxon reported that the basic elements of the most effective cost option were:

1 Close the boiler furnace doors and keep them closed and thereby increase average absorption efficiency of the boilers from 64% to 72% and save $16,000 per year.
Cost: $500 to install relay switches that prevent operation with the doors open.

2 Install 5 new burner cups and tune boiler to run at a single firing rate and 50% excess air. This will increase absorption efficiency from 72% to 80% and save $13,200 per year.
Cost: 1 or 2 man weeks of labor to repair dampers, replace cups, and tune boilers. Plus 2 weeks per year maintenance.

3 Install turbulators in boilers to reduce stack temperature by at least 100°F and cut fuel usage by 3.7%. This will save $4200 per year at a cost of $1750.

4 Remove handles from oil flow control valves in main boiler room and thereby assure that boiler stays in tune.

5 Install sequential draft regulators and dampers in the Annex and thereby eliminate the 15% loss incurred due to downtime draft pulling air through boiler. Savings are estimated to be $6100 per year. The cost would be $2000.

6 Repair or replace load management system in Annex to prevent tandem firing. Cost would be $500.

Thus with an investment of $5350 plus a healthy dose of Housing Authority man hours and operator training an annual savings of $39,500 can be realized.

Two observations on this report are of interest to the practicing architect. First, attention was focused exclusively on the boilers, and no consideration was given either to other parts of the heating system, such as the distribution system, or to the basis for heat loss in the structure itself. Given the problem of cutting fuel costs, the approach was to increase the efficiency of the boilers. The result was good as far as it went, but represented only a partial inquiry into the problem.

Second, the boiler investigation is the kind of technical investigation that is beyond the scope of most architects' knowledge. Indeed, many mechanical engineers are not equipped to conduct detailed boiler studies, and boiler specialists are needed for much of this work. As always, the architect's function is to see that the proper experts are brought in as required and to outline the scope of the work so that all aspects have been considered before final conclusions are drawn.

Chapter 15

Financing Conservation

Energy conservation is an emerging specialty whose track record is still being formed. Ordinarily, when an architect is retained by a client to design a building, the question of whether to build rarely arises because the client has already made this decision. The client decides which architect to hire, not whether one is needed. There is long historical precedent for the design/build process which results in a structure. The work is carried out against an established background of building codes, professional licensing procedures, and a whole set of routines which, worked out over a long period, provide the framework within which the industry operates.

Similar precedent governs the financing of buildings. Whether it be a bond issue to cover public works or a mortgage from a lending institution, both the funds and the conditions under which they are made available operate under a set of rules familiar to all the participants.

In time it is to be hoped that energy conservation activity will enjoy a similar framework; but that time is not yet at hand. Instead, work in this field is in a continually changing state of transition, and every project must make a case for itself and then develop its own financing.

When a traditional school is to be designed, for example, the work is carried out against a background of a host of similar, previously built structures. On the other hand, when an energy-conscious design is prepared for a school building, it is not yet possible to point to a host of similar designs which have already proved themselves and can serve as models for conservation activity. It takes an act of faith on the part of the client to authorize this work, and a similar act of faith on the part of the architect that in fact it is possible to deliver the energy savings that the participants hope to achieve. In time, an inventory of successfully designed structures which have proved themselves operationally will be built up, but in the interim we must rely primarily on professional analysis and intuition.

During this transition period, the architect's practice is further complicated by the fact that the architect's office itself is undergoing a learning process. It finds itself exploring many possibilities, only some of which prove feasible; and during the study large amounts of office time may be expended. Yet many clients feel that they should not have to pay extra for this education and that whatever is needed to produce an energy-efficient design should be included as part of the architect's basic services. After all, the client receives a rigorous structural analysis or a fully worked-out set of construction documents as part of the regular design package. Shouldn't energy conservation be included as well?

FEDERAL ACTIVITIES

It is against this background that the federal government has taken a number of steps to stimulate activity which might otherwise take a very long time to happen. Among these measures are:

- Grants
- Demonstration projects
- Energy budgets and codes
- Tax incentives
- Loan guarantees
- Energy audits
- Building programs

Grants

Several federal agencies sponsor applied research and development grants. These agencies include the Department of Energy, Federal Energy Administration, Department of Housing and Urban Development, National Science Foundation, General Services Administration, and NASA. The chief source of information about pending grants is the "Commerce Business Daily" published by the Department of Commerce.

Demonstration Projects

The funding sources for this work are essentially the same as for grants. Demonstration projects provide for actual construction of prototypes which, after postconstruction monitoring, can provide models for the building industry's future designs.

Energy Budgets and Codes

Regulations governing energy aspects of design fall into two categories. The first involves the promulgation of energy budgets under which an upper limit is set for annual energy consumption; this is generally expressed in British thermal units per square foot. This is a performance-type regulation in that the designer is given a goal to reach, but is not told how to achieve it. By contrast, energy codes (most of which incorporate ASHRAE's 90-75) are of the specification type, which detail how different parts of the building are to be constructed to meet thermal criteria.

Tax Incentives

The National Energy Act of 1978 provides tax incentives for energy conservation work for both commercial buildings (via investment tax credits) and individual residences. In the case of the residences, for example, there is a 15 percent credit, up to a maximum of $300, for money invested in insulation, caulking, storm windows and doors, weatherstripping, and the like. Additionally there is a 30 percent credit for the first $2000 and a 20 percent credit for the next $8000 spent on such renewable energy as solar, wind, and geothermal systems. The maximum credit for renewable energy work is $2200.

The impact of federal credits will be even greater in those states (California, Arizona) which have already been offering tax credits for various forms of energy activity.

Loan Guarantees

Various loan guarantee programs have been suggested as a way of financing energy conservation. In principle, the federal government would guarantee lending institutions against loss in fashion similar to the way FHA has insured and guaranteed home mortgages—with losses made up out of a kitty financed by insurance premiums. Thus, the Energy Conservation and Production Act (P.L. 94-385) contained provisions for $2 billion in 90 percent federal loan guarantees to retrofit various categories of buildings—commercial, small businesses, nonprofit, public, etc. Each borrower was limited to $5 million, and the guarantee could not exceed 25 percent of the fair market value of the building to be modified. This guarantee program was not implemented pending resolution of the long-delayed passage of the National Energy Act of 1978.

Energy Audits

As defined by the FEA, an energy audit means "a process . . . which identifies and specifies the energy and cost savings which are likely to be realized through the purchase and installation of an energy measure" (i.e., an energy conservation measure or a renewable resource energy measure). There are two types of FEA audits: information audits for use in state energy programs and verification audits for identifying and quantifying energy and cost savings as a result of particular retrofit measures.

The information audit has three subcategories: types A, B, and C. The type A audit involves a site visit by a qualified auditor, who advises building owners of likely energy conservation measures to take. Type B and C audits are oriented toward do-it-yourself building owners and homeowners. Qualifications of auditors are established by each state, most of whom permit architects and engineers to participate subject to certain qualifying procedures. Fees for these audits are also set by these states, but in general approximate 10 percent of the building's total energy costs for the preceding 12 months, when such energy costs were less than $40,000; $4000 or 5 percent of the building's total energy costs for the preceding 12 months, when such energy costs were between $40,000 and $1 million; $50,000 or 2.5 percent of the building's total energy costs for the preceding 12 months, when such energy costs exceeded $1 million.

Verification audits are detailed analyses of specific retrofit projects. The audit contains an analysis of the one or more building systems or industrial processes to be affected by the purchase and installation of a modification. A specific set of contents, format, and calculations is required, together with paybacks calculated on the basis of FEA price projections. Qualifications of verification auditors and allowable fees are similar to those for information auditors except that qualifying engineering background is heavier.

BUILDING OWNER AND OPERATOR FUNDING (INSTITUTIONAL)*

The elimination of energy cost, or cost avoidance, is an annually recurring savings that allows an institution to use financial resources that otherwise would have been lost up the smokestack.

Specific energy consumption reduction measures and modifications for an institution depend on its size and complexity. Regardless of these factors, however, there will be a need to seek capital for certain modifications, and it will be necessary to make cost-effective use of this capital. Seeking funds and using them effectively will vary among institu-

*The following section is reprinted from *Energy Management for Colleges and Universities* by permission of the National Association of College and University Business Offices.

tions, but there are some common energy finance and investment techniques that can be of assistance to every institution.

As the range of energy conservation measures and their respective costs are examined, one should keep in mind that energy management is an evolutionary process. The focus should first be on those measures that appear feasible within the level of staff and in-house expertise available. The institution should not attempt to accomplish, or implement, all options immediately, but should move deliberately and carefully, choosing those measures that have the greatest impact and that can be effectively handled by the existing organization. This step-by-step approach will ensure greater control over energy modifications and will assist in building the credibility needed to stimulate further funding of an energy management program.

Seeking Funds outside the Institution

As a first step, potential sources of funds for an energy management program should be listed. Discretionary funds that may be available for general upkeep and maintenance, as well as sources of capital funds for major refit and systems convert undertakings, should be investigated. The type of information that will be helpful for each source of funds includes:

- Name of source and person in charge
- Types of programs funded
- Types of funds disbursed (grants, loans, etc.)
- Restrictions imposed
- Time requirements
- Format of request
- Data required to support request

Questions on external sources can probably be answered by the agency or foundation involved. When possible, a few samples of previously approved proposals or budget requests should be obtained in order to have a better idea of what the agency considers a good request. Once a list of the sources of funds has been prepared, it should be reviewed. One should ask: Have all possible sources of funds been listed, including public agencies and private foundations? . . . Professional and technical journals may provide information on sources of funds.

In short, every possible source of present and future funds should be investigated, and comprehensive information should be kept on all sources with which there is regular communication. The institution's development (advancement) office is a valuable resource in the search for external sources of funds, both private and public, including alumni who may be interested in the program.

Seeking Funds within the Institution

Three types of situations are covered below: (1) institutions both public and private with quasi-endowment funds and/or free capital, (2) private institutions with no quasi-endowment funds or free capital, and (3) public institutions with no quasi-endowment funds or free capital.

In the case of institutions with quasi-endowment funds and/or free capital, the energy investment decision is usually not complicated. If the proposed energy modification earns a greater rate of return than would be realized in the stock or bond markets, the measure is probably worth implementing. However, the problem in securing capital for such modifications in these institutions is complicated by the competition for free capital dollars.

For example, the type of case that could be made in such a competitive atmosphere is that investment in a new library requires initial cost plus continuing costs for maintenance and operations. To be able to afford these new costs without additional revenue, the institution must reduce its annual operating costs. One of the best ways of reducing is investment of funds in energy conservation. Thus, when capital funds for the new library are available, the library's operating costs may be considered to be essentially "free" of an annual operating basis.

In the case of private institutions with no quasi-endowment funds or free capital, obtaining funds is more difficult. Although investment opportunities may be obvious, there may simply be no money available to invest. There are, however, several possible alternatives.

The first alternative is reduction of the operating budget and investment of "cut" dollars in energy improvements. Ideally, it could be stated that such a measure will result in a double savings, since the operating budget will have been cut and a return on energy investments will have been realized. The feasibility and effect of such a move must be carefully evaluated by an institution.

A second, somewhat unusual alternative is an energy conservation package for a prospective donor. Historically, it has been difficult to convince donors to give money for physical plant operations, but such donations could be viewed in a different light. For example, a donor may give $1 million for a new building, and by doing so creates new annual operating costs for the institution; however, if the same donor could be convinced to give the $1 million for energy conservation modifications that could reduce annual costs by $500,000, the donor would have effectively given $5 million worth of endowment money, using a rate of return of 10%.

In the case of public institutions, the problems of acquiring funds are usually more complex than those faced by private institutions. Legislatures or other public bodies may be reluctant to increase budgets, even when a clear payback is evident; and if they are willing to do so, there may be strong pressure to spend these increases on new ventures rather than on long-term savings.

The economic arguments for energy investments by public institutions are the same as those for private institutions. Investment in energy conservation projects produces long-term savings for the institution and the taxpayer. However, the politics of public institutions are complex, and in some cases a rigid approach to budgeting may make it difficult to progress. It is necessary to examine specific approval processes and to identify those persons needing the most persuasion, and then to determine how that "persuasion" can best be presented.

Resistance may come from the president of an institution because he or she is feeling pressure from a variety of departments for funds, and thus energy conservation expenditures may be viewed as of relatively low priority. A possible solution to this type of problem may be to plan an energy education program so that the academic community can understand the long-term benefits that such a program will bring.

BUILDING OWNER AND OPERATOR FUNDING (PRIVATE OWNERSHIP)

As architects who have worked for both know only too well, there is a vast difference between having an institution for a client and working for private owners (sometimes referred to as either investment or speculative builders).

The private owner is almost always cash-poor, since whatever capital is available is generally quickly invested in highly leveraged situations. Further, since most builders tend to specialize in one or two particular types of buildings, such as office buildings or high-rise housing, and since they tend to be involved with all phases of planning, building, and then operating the structure, they inevitably possess greatly detailed knowledge of their buildings.

With respect to energy conservation, even though it is an emerging technique with a whole new set of rules, the private owner's confidence is extended to cover this new field. Accordingly, the private owner tends to be a "tough sell" and usually establishes a scenario in which the architect's brains get picked clean but the architect rarely ends up with a job. During the interview, the owner tries to pick up as many bits of advice as possible concerning quick-fix items which can be accomplished with little or no investment. Usually the owner is unwilling to pay for having the building checked out thoroughly (after all, who knows it best of all?) and is rarely interested in anything with a payback of greater than a year.

The exception to this rather strained situation is the case where the owner has had a continuing relationship with the architect over many years. In the author's experience, this personal relationship can often outweigh the habits by which the builder ordinarily lives and can result in commissions which are mutually profitable to both. It is, however, extremely difficult to convince a builder whom one has met for the first time.

As time goes by, two events will change this situation for the better. First, the benefits of energy conservation will slowly produce a track record of which everyone will want to take advantage; second, financing, probably through a system of loan guarantees by the government, will furnish a source of capital to the builder and avoid the necessity of having to dig into her or his own pocket.

QUASI-PUBLIC AGENCY FUNDING

Still on the horizon but receiving increasing attention is the possibility of having quasi-public agencies provide the capital needed for energy work, including conservation. These would be similar to the many authorities established across the country to take care of specialized needs while attempting to avoid the cumbersome bureaucracies of government.

Typical is the proposed Energy Corporation of the Northeast (ENCONO). Its purpose is to plan, expedite, guarantee, and finance projects related to reducing the costs, or increasing the supply, of energy in the Northeast United States.

The planning function could include the identification of energy problems and the evaluation of energy costs and consumption in relation to economic development. The expediting function could involve the corporation acting as an intermediary with all levels of government and speeding the process of regulatory approvals for corporation projects. Projects eligible for corporation financial support include the development and marketing of energy conservation techniques.

Generally, ENCONO would provide a means for accelerating energy conservation activities through large-scale regional marketing efforts, the coordination of material and financing elements, and the effective interchange of information between various organizations.

Such activities, for example, would include National Energy Plan measures, such as business and residential retrofit tax credits, solar incentives, utility home energy conservation program, and the schools and hospital retrofit programs, as well as financial assistance programs authorized by the Energy Policy and Conservation Act of 1975 and the Energy Conservation and Production Act of 1976.

ENCONO hopes to play a catalytic role by participating with capital and guarantees, in combination with banks and utilities, in major building retrofit projects; designing and implementing, in coordination with the states, a regional marketing program; and providing, on a selective basis, financial assistance to regionally focused energy conservation research, development, and demonstration projects.

Bibliography

BOOKS

Anderson, Bruce: *Solar Energy—Fundamentals in Building Design,* McGraw-Hill, New York, 1977.
Burberry, Peter: *Building for Energy Conservation,* Wiley, New York, 1978.
Coonley, Douglas R.: *An Introduction to the Use of Wind,* Total Environmental Action, Harrisville, New Hampshire, 1975.
Dubin, Fred S.: *Energy Conservation Standards,* McGraw-Hill, New York, 1978.
Fitch, James Marston: *American Building,* Schocken Books, New York, 1975.
Griffin, C. W.: *Energy Conservation in Buildings,* The Construction Specifications Institute Inc., 1974.
Kendrick, Lee: *Energy Conservation and Management for Buildings,* The Office of Lee Kendrick, Falls Church, Virginia, 1977.
Kreider, Jan, and Frank Kreith: *Solar Heating and Cooling,* McGraw-Hill, New York, 1975.
Mazria, Edward: *The Passive Solar Energy Book,* Rodale Press, Eammaus, Pennsylvania, 1979.
McGuinness, William J., and Benjamin Stein: *Mechanical and Electrical Equipment for Buildings,* Wiley, New York, 1971.
Shurcliff, W. A.: *Solar Heated Buildings,* W. A. Shurcliff, Cambridge, Massachusetts, 1977.
Stein, Richard G.: *Architecture and Energy,* Anchor Press/Doubleday, New York, 1977.
Watson, Donald: *Energy Conservation in Building Design,* McGraw-Hill, New York, 1979.

PUBLICATIONS OF ASSOCIATIONS, ORGANIZATIONS, AND BUSINESSES

ASHRAE Handbook and Product Directory, ASHRAE; *Applications,* 1978; *Fundamentals,* 1977; *Systems,* 1976; *Equipment,* 1975.
ASHRAE Standards 90-75, 100.1P, 100.2P, 100.3P, and 100.5P, ASHRAE 1977.
"Conservation, Cleanliness and Coal," General Electric Research and Development Center, 1978.
"Design Considerations for Solar Systems," Grumman Energy Systems, Inc., 1978.
"The Economy of Energy Conservation in Educational Facilities," Educational Facilities Laboratories, 1973.
"Energy and the Built Environment," AIA, 1974.

"Energy Conservation in Building Design," AIA Research Corporation, 1976.
"Energy Conscious Design of New Buildings," AIA Research Corporation, 1976.
"Energy Conservation with Comfort," Honeywell, 1976.
"Energy Consumption for a Typical Office Building," WTA Computer Services Inc., 1974.
"Energy Cost Reduction for Apartment Owners and Managers," Institute Of Real Estate Management, 1977.
"Energy Cost Reduction in Retailing," American Retail Federation, 1977.
"Energy Management for Colleges and Universities," Energy Task Force, 1977.
"Energy Notebook Information Service," AIA, 1977 and 1978.
"Energy—The Case for Conservation," Worldwatch Institute, 1976.
"Energy Use, Conservation and Supply," American Association for the Advancement of Science, 1974.
"Evaluation of Energy Consumption in Office Buildings," Enviro-Management and Research, Inc., 1975.
"Impact Assessment of ASHRAE Standard 90-75," Arthur D. Little, Inc., 1976.
"Industrial Energy Conservation," General Motors Corporation, 1977.
"Life Cycle Cost Analysis," AIA, 1977.
"Modern Utilization of Infrared Technology," The Society of Photo-Optical Instrumentation Engineers, 1977.
"A Nation of Energy Efficient Buildings by 1990," AIA, 1974.
"A Plan for Total Energy Efficiency in Buildings," AIA, 1975.
"Practical Energy Management in Health Care Institutions," Blue Cross of Greater Philadelphia, 1977.
Proceedings of the 1977 Annual Meeting, American Section of ISES, 1977.
Proceedings of the 2d National Passive Solar Conference, American Section of ISES, 1978.
Proceedings of the 12th Intersociety Energy Conversion Engineering Conference, American Nuclear Society, 1977.
"Skylite Annual Energy Balance," Architectural Aluminum Manufacturers Association, 1977.
"Solar Dwelling Design Concepts," AIA Research Corporation, 1976.
"Solar Energy Report," Sutro and Co., 1977.
"Solar Heated Houses for New England," AIA Research Corporation, 1975.
"State Energy Conservation Plans," AIA, 1977.
"States' Energy Conservation Standards," National Institute of Building Sciences, 1978.
"Thermal Performance Guidelines for One and Two Family Dwellings," National Association of Homebuilders, 1977.
"Total Energy Management," National Electrical Contractors Association, 1976.
"Visio-Therm Economics," Solar-X Corporation, 1976.

PERIODICALS

Energy & The Builder: Professional Builder Magazine, 1977.
Energy Users Report (weekly), The Bureau of National Affairs, Inc., 1977, 1978.
The 1978 Energy Guidebook, Power Magazine, 1978.
Windows for Energy Efficient Buildings, Lawrence Berkeley Laboratory, 1979.

GOVERNMENT PUBLICATIONS

"All Electric Homes in the United States," Federal Power Commission, 1976.
"Building Energy Handbook," Department of Commerce, 1976.

"Buying Solar," Federal Energy Administration, 1976.

"Conversion to Coal," New York State Legislative Commission on Energy Systems, 1978.

"Designing an Energy-Efficient Building," General Services Administration, 1975.

"Economic Thickness for Industrial Insulation," Federal Energy Administration, 1976.

"An Energy Action Plan for Small Commercial Buildings," New York State Energy Office, 1978.

"Energy and the Environment," Energy Research & Development Administration, 1975.

"Energy Conservation Applied to Office Lighting," Department of Commerce, 1975.

"Energy Conservation in the Commercial Sector," Federal Energy Administration, 1977.

"Energy Conservation Design Guidelines for New Office Buildings," General Services Administration, 1977.

"Energy Conservation Guidelines for Existing Office Buildings," General Services Administration, 1975.

"Energy Conservation Handbook for Light Industries and Commercial Buildings," Department of Commerce, 1974.

"Energy Conservation Handbook for Public and Commercial Buildings," Intergovernmental Planning Coordinating Committee, 1976.

"Energy Conservation and Window Systems," Department of Commerce, 1975.

"Energy Cost Reduction for Gasoline Service Stations," Federal Energy Administration, 1977.

"Energy Management and Economic Sense for Retailers," Department of Commerce, 1974.

"An Energy Management Program for Multi-Story Office Buildings," New York State Energy Office, 1978.

"Energy Utilization Index Method," Department of Commerce, 1977.

"Evaluation of Ventilation and Energy in Existing Schools," National Bureau of Standards, 1977.

"Financing Energy Development," Congressional Budget Office, 1976.

"Forms for Energy Survey and Appraisal," Department of Commerce, 1976.

"A Forum on Solar Access," National Solar Heating and Cooling Information Center, 1977.

"Geothermal Energy Program," National Science Foundation, 1976.

"Guidelines for Saving Energy in Existing Buildings," Federal Energy Administration, 1975.

"Guide to Energy Conservation for Food Service," Federal Energy Administration, 1977.

"Guide to Energy Conservation for Grocery Stores," Federal Energy Administration, 1977.

"Handling Fuel and Fuel Problems," Federal Energy Administration, 1975.

"Hourly Solar Radiation Data," Department of Commerce, 1977.

"How to Save Energy and Money in Your Home," New York State Energy Office, 1978.

" A Human Factors/Systems Viewpoint," Department of Commerce, 1976.

"Incentives for Energy Conservation in Multi-Family Housing," Department of Commerce, 1975.

"In the Bank or up the Chimney?" Department of Housing and Urban Development, 1975.

"Lighting and Thermal Operations," Federal Energy Administration, 1975.

"Low-Temperature Thermal Energy Storage Program," Oak Ridge National Laboratory, 1979.

"Making the Most of Your Energy Dollars," National Bureau of Standards, 1975.

"Minimum Energy Dwelling (MED) Workbook," Department of Energy, 1977.

"National Energy Outlook," Federal Energy Administration, 1976.

"A National Plan for Energy Research," Energy Research & Development Administration, 1976.

"Passive Design Ideas for the Energy Conscious Architect," The National Solar Heating and Cooling Information Center, 1978.

Proceedings of the 2d Annual Thermal Energy Storage Exchange, Oak Ridge National Laboratory, 1977.

Proceedings of the Workshop on Solar Collectors, National Science Foundation, 1974.
"Project Retro-Tech Teachers' Kit," Department of Commerce, 1975.
"Regulations Establishing Energy Conservation Standards," California Energy Resources Commission, 1977.
"Site Energy Handbook," Department of Commerce, 1976.
"Site Visit Report" Federal Energy Administration, 1976.
"State Energy Conservation Construction Code," State of New York, 1979.
"Total Energy Management for Hospitals," Department of Health, Education, and Welfare, 1977.
"Total Energy Management for Nursing Homes," Department of Health, Education, and Welfare, 1977.
"The Use of Solar Energy for Space Heating and Hot Water," Massachusetts Energy Policy Office, 1976.
"Water for Energy Self-Sufficiency," Water Resources Council, 1974.
"Ways to Conserve Energy in Your Business," Department of Commerce, 1974.
"Window Design Strategies to Conserve Energy," Department of Commerce, 1977.

Index